COMPLETE BOOK ON
WASTE WATER TREATMENT

COMPLETE BOOK ON
WASTE WATER TREATMENT

By :

Maryam Noor

2007

CYBER TECH PUBLICATIONS

4264/3, Ansari Road, Darya Ganj
New Delhi 110002 Ph. : 23244078

COMPLETE BOOK ON WASTE WATER TREATMENT

Maryam Noor

First Edition 2007

Published by :

G.S. Rawat for **Cyber Tech Publications**
4264/3, Ansari Road, Darya Ganj, New Delhi 110002

Printed at :

PREFACE

Domestic sewage is a mixture of waste waters arising from household waste facilities including toilets, bathrooms and kitchens. It contains pathogenic microorganisms (bacteria and viruses that cause disease) and creates odour nuisance (sulphurous smell). Most sewage-treatment plants are designed to purify domestic sewage to an effluent quality adequate for discharge to the nearest watercourse. An exception exists in arid climates where waste waters are often used for landscape irrigation and other industrial and agricultural purposes.

Sewage is typically about 99.9 per cent water; the other 0.1 per cent is mainly organic materials (organic means derived from living things), a wide range of compounds comprising the elements carbon, hydrogen and oxygen plus smaller amounts of nitrogen and sulphur. The organic impurities enter the water during its various household uses.

—Author

CONTENTS

1

PREMILINARY TREATMENT OF WASTE WATER

WHY TREAT WATER?

Water must be treated every time one of the analytical measurements rises above the current legal standard in the country in question. WHO (World Health Organization) sets recommendations for each measurement that must be followed by each country depending upon health conditions in that country and the state of its economy; the goal is to establish national standards. All water made available to consumers through a supply system must be treated even if only a small fraction of it is directly consumed by the individual. It would be hazardous to public health and economically prohibitive to set up a double supply system with one distributing water for consumption and the other supplying water for other uses. Water coming out of the consumer's faucet must be "potable", that is, it must conform to standards in force regardless of how the consumer chooses to use it.

The unsophisticated but essential methods of removing detritus and grit from sewage and other waste waters. Detritus, as used in waste water engineering, is a collective team for gross solids, rag, plastics and general trash, which is separately collected before grit removal at a conventional sewage treatment works (STW); the methods and equipment are equally useful for STW and industrial waste water treatment.

Grit comprises inorganic material ranging from fine sand/silt particles about 0.1 mm in diameter to material over 1 cm^3 in volume and of irregular shape. Surface water run-off and general erosion of the sewerage system are the usual sources. Screens to remove detritus vary from coarse (more than 50mm between the bars) to fine (1—2mm), and are usually electrically driven. There are a number of manufacturers and designs, all with their own characteristics.

Simple filter sacks have recently become available, and provide an effective solution for screening out plastics, which are a particular contemporary problem in that they break through screens to primary sedimentation tanks and secondary treatment, and can even penetrate to the works outfall. Wire baskets and grids more suited to the cleaner factory effluent can often be made in-house out of cheap weld mesh. Grit is highly abrasive and often blocks pipework; separation is based on reducing flow rates to assist gravity settlement, although the material at a sewage works is too variable in grain size and organic content to have any value, and is dumped or disposed to landfill.

Physical methods of separating oil, petrol and fuel/water mixtures are also described and all are simple means of avoiding a highly visible and unattractive type of pollution, one which is immediately obvious to the general public and the regulating bodies.

Industrial effluents containing oil emulsions require more specialised treatment by cracking with acid, or solvent extraction. They are often successfully handled in admixture with other organic wastes by one of the biological processes.

THE IMPORTANCE OF PRELIMINARY TREATMENT AND INITIAL CONSIDERATIONS

Preliminary treatment is essential at STWs as the water companies have very limited control over the amount of general rubbish that can end up in sewers, and even less over the grit and inorganic particles arising from erosion of pipes and joints and from roads and paths where the sewerage system is combined.

Mechanical coarse screens are also located at some pumping stations, particularly on combined (*i.e.*, foul and surface) water sewerage systems, and in drainage systems with a number of pumping stages, those suffering considerable infiltration or where rag and large items are often found in the sewage. These factors are always worth considering in any new scheme or rehabilitation. Pump blockage and damage in the sewerage system is expensive on parts and manpower, particularly overtime, and there is the associated risk of flooding and pollution. By contrast, even a poorly run industrial site is in a much better position to prevent gross material from entering liquid effluent flow by simple housekeeping, and strategically placed grids and wire baskets.

From an economical viewpoint, the methods and formulae used by the water companies to charge for trade effluent always include a 'strength' factor. Some form of screening will often reduce the overall effluent strength by sifting out larger solids before they break down and require removal by sedimentation or biological means.

If an effluent plant is operating on the limit of organic loading and performance, fine screening can tip the balance away from the overloaded state, or at least give some breathing space. Hydraulic loading and head loss across screens are attendant factors to consider in this situation.

Satisfactory operation of all subsequent stages of a treatment plant often critically depends on correct installation, operation and regular maintenance of screens, traps and interceptors. A number of serious problems are likely to develop from neglecting these devices, including excessive wear and breakdown of pumps, blockage and erosion of pipework, and increased maintenance requirements to equipment throughout the plant. There is then the associated chance of the works effluent failing discharge Consent limits.

Accumulations of detritus reduce the efficiency of biological treatment by generating dead volume in tanks, and may impair settlement and sludge digestion rates, the latter often caused by rag 'balling'. Detritus building up on weirs causes erratic flow measurement, and influences plant operation by incorrect chemical conditioning dosing or incorrect recirculation rates. If pumped, the latter can induce wide flow variation through the plant which may also encourage departures of effluent standard outside Consent limits.

Screenings in sewage have changed substantially in recent years, consisting of far more non-biodegradable plastics material of fine gauge and fewer cellulose-based products (see CIR1A/WRC (1984)). Plastic is persistent, and many coarse screen designs retain little of it, while some older designs of macerators and shredders cut it into ribbons which are even more difficult to retain.

Substantial growth has taken place in the fine screen market during the 1980s, as plastics are now prohibited substances in the aquatic environment and must not be discharged. The trend has been to wash and remove screenings, followed by landfill disposal or incineration. Dewatering and compaction devices frequently complement a package screen installation, and substantially reduce screenings disposal volumes. There has been a parallel trend in Western Europe for the *per capita* volume of screenings to increase, and 0.3m^3/day per 1000 population is often quoted as an average value.

The waste disposal unit under the kitchen sink is the most likely culprit causing this trend, and whilst mainly increasing the biological and sludge loadings to the works, also generates fibrous material likely to be trapped and effectively blind fine screens below 5mm mesh size. The water industry might justifiably promote a policy of actively discouraging these devices which are estimated to increase the per capita solids loadings to the sewer by 25% per day.

Screenings and detritus, particularly of sewage origin, and oil/water mixtures are very noticeable near beaches or inland surface waters, and frequently attract attention by the public and regulatory bodies out of proportion to the actual pollution level because they are considered aesthetically revolting. The function of screening sewage discharges to sea is therefore largely concerned with reducing the identifiable nature of the solids, as much of the material tends to float.

There are a number of sewage discharges to sea and many worldwide which receive no treatment other than preliminary solids screening. Screening is too coarse at many of them, reliance being placed on dilution mechanical breakdown by waves and tide and distance from the shore. Many storm overflows are not screened at all. Guidelines issued by Her Majesty's Inspectorate of Pollution (HMIP) as a letter to the Water Authorities in 1988 suggest that a mesh size greater than 6mm is unlikely to achieve adequate performance of long sea outfalls which are the subject of EEC Directive 76/160/EEC (1976) Concerning the Quality of Bathing Waters: '...that all flows should be subject to fine screening...' The HMIP letter also discusses the extent of screening in relation to the magnitude of storm flows. Other aspects to consider for marine discharges include the general requirements of Directive 91/271/EEC (1991), the Council Directive Concerning Urban Wastewater Treatment, which dictates that preliminary screening must be continuously effective, and adds the requirement for primary settlement and possibly secondary aeration to many of these discharges.

In such cases, preliminary treatment might help to reduce the bacterial numbers in subsequent discharges by solids removal, also useful in complying with Directive 76/160/EEC (1976). 'Might' is the operative word, as the action of many screens is to break down solids as well as remove larger material so that there is no difference between the organic load measured as Biochemical Oxygen Demand (BOD) either side of some screen designs.

Discharges of sewage to the marine and estuarine environment are currently receiving particular attention in Europe. A useful reference paper on screen performance, particularly in relation to the plastics problem, is provided by Thomas *et al.* (1989).

Conventional grit removal on sewage works takes place after screening to remove detritus. Some grit will therefore be trapped in fine screens and will act as an abrasive on the screen surfaces and the cutters of macerators or comminutors. It is desirable, therefore, to specify hardened or stainless materials for screens.

Unremoved grit tends to accumulate on sewage works in sludge digestion and aeration tanks, reducing effective capacity, circulation and sludge retention times. This has several negative effects. These include reducing gas production rates and energy availability in boilers where external and expensive energy 'topping up' may become necessary; lowering the quality of the sludge and liquors produced, affecting works performance; and generating heavy maintenance costs on mixers and pumps arising from abrasive wear to seals, bearings and metalwork.

Oil discharges to treatment plant make an unsightly mess of tank and channel walls which are labour intensive to clean; the oil also coats filter surfaces, reducing oxygen transfer and performance. There is frequently an odour problem from volatile components or if the material emulsifies and turns rancid, and many mineral oils contain toxic organics and heavy metals after use. Trade effluent Consents specifically exclude flammable and explosive fuels from discharges to the foul sewer, and contemporary limits of 500mg/l for oil are typical from garage forecourt interceptors and car washes.

It is worth noting that oil discharges are the easiest to trace back to source either in rivers or sewers, by virtue of their persistence and ready observation, and the elegantly conclusive analytical techniques which are able to uniquely 'fingerprint' oil; thus a number of cases brought to court by the National Rivers Authority (NRA, 1991) have resulted in successful prosecutions and the imposition of spectacular fines. There is thus practical, commercial and legal sense behind some of the humble devices now described.

CRITERIA FOR CHOICE

A number of factors determines which water should be treated before it is distributed. For each available water source (groundwater, flowing or stored surface water) we must assess:

- the **quantity:** the "source" must be able to furnish the required quantity of water at all times. In countries where the amount of rainfall varies greatly, it may be necessary to set up a reservoir and fill it during the rainy season with enough water to meet the needs of the dry season,
- the **quality:** in some countries the quality of available raw water must meet current legal standards. A determination must be made as to which treatment process would be best to make the water drinkable. This process would have to be weighed against previous years' fluctuations in the quality of water (daily, seasonal, climate variations) and against potential fluctuations that may occur in the future (with the construction of a dam, for example).

  In addition, it must be remembered that groundwater is not synonymous with pure water: actually, many aquifers are polluted with bacteria, nitrates, phytosanitary products, chlorinated solvents or hydrocarbons,

- the **cost:** for each available source, the capital and operating costs must be compared so as to assure both quantity and quality of water for distribution: storage and transport of raw water, water treatment, storage and transport of treated water.

BIOLOGICAL IMPURITIES

All water is vulnerable to pollution by microorganisms,

Bacteria, and Viruses

The bacteria which are indicators of fecal contamination are brought into the environment by the discharge of municipal wastewater, which may or may not have been treated. These bacteria indicate possible contamination by bacteria or viruses which are pathogenic for man.

The growth of common germs may create major problems in the distribution system: consumption of dissolved oxygen, corrosion, bad taste of water.

Various Microorganisms Phytoplankton and Zooplankton

Surface water contains many organisms that make up phytoplankton and zooplankton. Many of these organisms, such as Acrinomyceaes and Cyanophyceae, secrete compounds (such as geosmin) which give water a bad taste and smell. Others are pathogenic for man (amoebae).

The presence of algae and macroorganisms in the water system (Asellus, Copepoda, Nematoda) is unpleasant for the consumer and their growth can lead to major disturbances (proliferation, sediments, appearance of anaerobic conditions),

Moreover, while they grow or at their death, some microorganisms (Cyanophyceae, for instance), excrete metabolites that are toxic to higher animals.

MINERAL IMPURITIES

Some of these impurities influence the organoleptic qualities of water, its aesthetic appearance or its behaviour in the distribution system but do not have any appreciable effect on the health of the consumer, whereas others have a known effect.

Impurities which do not have Any Appreciable Effect on Health

- **Turbidity**

Together with colour, this is the first quality the consumer notices. All water is turbid bur if the turbidity is too noticeable, the consumer rejects the water.

There are other reasons, also, why turbidity must be removed:

- ◆ to permit proper disinfection of water,
- ◆ to remove any pollutant adsorbed on suspended solids (heavy metals, etc.),
- ◆ to guard against sedimentation in the distribution system.

- **Colour**

Colour may be due to some mineral impurities (iron, etc.), but it may equally be due to some organic matter (humic and fulvic acids). It must be removed in order to make the water pleasant to drink. The removal of colour goes hand in hand with the removal of some undesirable organic substances (precursors of haloforms, etc.).

- **Mineralization**

Alkalinity and hardness contribute to the carbonate balance of water together with the pH and the dissolved carbonic acid. An attempt is made to attain such a balance in water in order to avoid scale formation in, or corrosion of, the system. If the amount of sulphates in water is too high, it affects the taste and gives it the quality of a laxative. If the amount of chlorides is too high, it also affects the taste and the water becomes corrosive.

- **Certain Metals**

Iron and manganese can colour the water and initiate sedimentation in the system. This can result in corrosion. In addition, they affect the organoleptic quality of water as do other metals: copper, aluminium, zinc.

- **Dissolved Gases**

H_2S indicates an anaerobic condition and an oxidation-reduction potential that is too low; it causes a bad smell and may initiate corrosion. It must be removed.

- **Ammonium NH_4^+**

This has no appreciable effect on the health of the consumer but its presence in water indicates pollution. In deep-lying water, NH_4 may be equally due to reducing conditions in an aquifer. Ammonium must be removed from water for consumption as it is a nutritive element that may allow some bacteria to grow in distribution systems.

Impurities Affecting Health

Study Methods

It is difficult ro study the effect of a given product directly on man. We can guard against accidents from acute toxicity by certain products which lead to death (poisoning). In the same measure, we can carry out epidemiological studies which correlate the ingestion of certain products with death, cancer risk or a given illness.

However, given the number of factors that can enter into the environment of each person along with the mobility of populations in the modem world, epidemiological studies are long, costly and the results are questionable. It is therefore a preferable to employ experimental methods. To gain a better understanding of the effects of various pollutants on the health of consumers, studies and experiments are not performed directly on man but rather on those animals whose reactions are known to be similar to man's. The results obtained are then extrapolated to conform to man by means of models which best reproduce the transfer from animal to man.

The effects observed on different animals can be described in various ways:

- **acute toxicity:** the product quickly leads ro death of the animal. The LD 50 value indicates the lethal dose leading ro death in 50% of individuals in a given time (24 hours, for example),
- **chronic toxicity:** this is a dose, which when ingested on a daily basis, leads to premature death of individuals. ADI (Acceptable Daily Intake) indicates a maximum dose ingested daily over a lifetime that the metabolism of an individual is able to withstand without risk,
- **cytotoxicity:** toxicity may be studied by using cell cultures in the place of living organisms: the product under study leads to the death of a certain percentage of cells: this defines, then, cytotoxicity,
- **mutagenicity:** ingesting a product causes mutations in individuals. The risk of mutagenicity exists regardless of the dose ingested. The risk is low if the ingested dose is low and the risk increases if the ingested dose is increased,
- **carcinogenic effect:** exposing an individual to the product under study or having him ingest it causes a malignant rumor to appear. As for mutagenicity, the carcinogenic effect exists whatever the dose ingested.

Chief Impurities

- **Metals**

In particular, cadmium, chromium, lead, mercury, selenium and arsenic must be removed from water. Generally, they are adsorbed on suspended solids in raw water and the removal of these suspended solids is therefore sufficient to ensure their removal. In some cases, the metals may be complexes either of natural organic substances (mercury, for example), or of chemical compounds discharged by industry or homes. The treatment must be able to destroy these complexes to ensure their removal.

- **Nitrates**

There has been a general increase in the concentration of nitrates in raw water and these must be removed because they are known to have a harmful effect when present in high concentrations in drinking water. They cause methaemo-globinaemia in infants.

- **Asbestos Fibres**

Although asbestos has been recognized as carcinogenic when inhaled from the air, the carcinogenic effect of asbestos fibres in drinking water has not been clearly demonstrated. It is, however, desirable to remove them as completely as possible. Asbestos fibres can be carried in steam (boiling water, showers, etc.). By reducing turbidity, a reasonable amount of asbestos fibres can be removed.

- **Hardness**

It has no tangible effect on health. However, softening water by ion exchange in the sodium cycle leads to a higher content of sodium in the water; a high sodium content seems to promote hypertension.

- **Fluorine**

Too much fluorine leads to the fluorosis of bones. In such a case, the fluorine concentration must be reduced.

ORGANIC IMPURITIES

Many natural organic substances are found in ground- and surface water. These are classified in six main groups: humic substances, hydrophilic acids, carboxylic acids, peptides and amino acids, carbohydrates and hydrocarbons. Organic substances are defined analytically by measuring them against overall indices, reagent groups, or well-defined substances.

Overall Parameters

These are measurements which do not show the concentration of a well-defined organic substance, but rather indicate a characteristic common to several substances.

Permanganate Value and **Total Organic Carbon (TOC)** indicate the concentration of organic substances that may be natural (fulvic acid, humic acid) or that come from municipal or industrial effluents. The concentration of these parameters should be reduced as much as possible in treated water to avoid the growth of microorganisms in the distribution system

Absorbance measured in ultraviolet rays at 254 nm indicates the concentration of double bonds (aliphatic, carboxylic, benzenoid).

The **TOCL (or TOX)** indicates the concentration of chlorinated (or halogenated) organic substances. It increases following chlorination of water and should be as low as possible by the time it reaches the consumer's faucet.

◆ The usefulness of overall parameters

It is impossible to relate an overall parameter to a measurement of toxicity and/or cytotoxicity or carcinogenic and/or mutagenic effect; and thus it is impossible to determine concentration limits based on some effect or other on the health of the consumer. However, being aware of the overall parameters during the whole treatment procedure allows us to make optimum use of a treatment plant and compare procedures.

Pesticides and Plant Growth Regulators

These products are used in agriculture to combat organisms that attack vegetables. Some become hydrolysed rather quickly but others are especially resistant and accumulate in the food chain. Some of these products are toxic while others are mutagenous or carcinogenic. The WHO issued (1985) the ADI for some products.

Simazin and atrazine are now seen more and more frequently. These dangerous products must be removed as completely as possible to protect the health of the consumer.

Table 1

Compound or group of isomers	Recommendations $\mu g.l^{-1}$	ADI mg per kg of the person
DDT (all isomers)	1	0.005
Aldrin and dialdrin	0.03	0.0001
Chlordan (all isomers)	0.5	0.001
Heptachlor and hexachloro-epoxy	0.1	0.0005
Gamma HCH (hexachlorohexane-lindane)	3	0.01
Methoxychlor	30	0.1
2, 4-dichlorophenoxyacetic acid	100	0.3

Chlorinated Solvents

These substances have contaminated many deep-lying waters, through industrial discharges, dumping into disused wells and leachates. These substances are carcinogenic or mutagenic and it is important to remove them. Notable are: carbon tetrachloride, 1, 2-dichloroethane, 1, 1-dichloroethene and 1, 1, 1-trichloroethane.

Phenols and Derivatives

Phenols and their derivatives are the mark of industrial pollution. Their worst effect is that, in the presence of chlorine, very small quantities of these products leave a taste of chlorophenol. Normally there is no taste if the pure phenol content is kept down to 1 $\mu g.l^{-1}$ but there is sometimes a slight taste of chlorophenol with a content of 0.1 to 0.01$\mu g.l^{-1}$.

When chlorophenols are first detected organoleptically, they are still far below the threshold where they pose a danger to the health of the consumer; these substances must be removed until they can no longer be tasted.

Hydrocarbons

The hydrocarbons likely to pollute surface or groundwater supplies come mainly from oil refinery wastes, industrial effluents of various kinds, gasworks effluents, fumes, etc. Aromatic hydrocarbons are especially soluble.

Such wastes may contain petroleum, kerosene, petrol, fuel oil, other oils or lubricants.

Biodegradability is slow. Accidental pollution is short-lived at the intake of a river purification plant but can last a long time in the case of groundwater (up to several years because of the soil's power of retention). This is why groundwater supplies have to be strictly protected against the risk of hydrocarbon contamination.

- **Harmful and Toxic Effects**

  - formation of a film which interferes with the re-oxygenation and self-purification of surface water,
  - interference with the operation of drinking water treatment plants; flocculation and settling are affected and the hydrocarbon is likely to remain in the filter material for a long time,

- the taste and smell threshold varies very widely according to the product involved (from 0.5mg.l^{-1} for petrol to 1 mg.l^{-1} for oils and lubricants),
- toxicity; there is a danger of toxicity in drinking water at concentrations above those at which taste and smell appear. Skin troubles have been caused by fuel oil additives.

Polycyclic Aromatic Hydrocarbons (PAH)

Some of these substances are very carcinogenic and it is important to remove them completely before the water is distributed. Noteworthy arbenzo(3, 4)pyrene,e, benzo (1 1,12)fluoranthene, benzo(l,12)perylene, benzo(3,4) fluoranthene and indeno(1,2,3-Cd)pyrene.

Polychlorobiphenyls (PCB)

These products have been developed over the past few years for various uses: plasticizers, solvents, lubricants, hydraulic heat transfer fluids. They are mainly used in the manufacture of transformers and capacitors. In France they are known by their brand name, Pyralene.

Through combustion or pyrolysis, they can give off products suspected to be very toxic: polychlorodibenzofurans ("furans") and polychlorodibenzodioxins ("dioxins"). These products are especially stable, and as such, are dispersed in the environment. They are assimilated by living organisms and may be carried in the food chain. Although it has not yet been formally demonstrated, their toxicity is strongly suspected.

Detergents

Detergents are synthetic surface-active compounds which enter the water with municipal and industrial effluents.

Commercial products contain active compounds in the form of surfactants and aids.

- **Surfactants:** their structure modifies the physical properties of surfaces by lowering surface tension and gives them cleaning power. There are various types:
    - **anionic surfactants:** for a long time the most commonly used substances were "hard," slightly biodegradable, branched- chain products, such as the alkylbenzene sulphonates (ABS), which have been mainly responsible for the problems created by the presence of detergents in water. They have usually been replaced by at least 80% biodegradable, linear-chain detergents.

      The concentration of anionic surfactants can be measured easily by methylene blue analysis; their biodegradability over a period can then be followed without difficulty,
    - non-ionic surfactants (those now used have an alkylphenol or even polyethoxyl alcohol base). These are being used on an increasing scale, but problems have still to be solved as regards dosing,
    - cationic surfactants, consisting of quaternary ammonium salts, are little used and are reserved for special uses linked with their biostatic properties.
- **Aids** : These include:
    - aids proper such as polyphosphates, carbonates, silicates,
    - chelating and complexing agents (polyphosphates),
    - reinforcing agents to improve the action of the active constituents (amino oxides, carboxymethylcellulose, alkanolamides),
    - **additives:** bleaching agents, perborates, optical bleaches, dyes, perfumes,

- ◆ mineral salt fillers to improve the appearance of products,
- ◆ enzymes which should be regarded as pre-adjuvants and help to hydrolyse certain types of fouling.

- **Concentrations Found in Water** : Before the introduction of biodegradable products, the concentration of anionic detergents in river water varied from 0.05 to 6 mg.l^{-1}. The figure has dropped since.

  Concentrations of non-ionic detergents are difficult to express because of the many methods of analysis, their accuracy and limits of detection.

- **Harmful Effects** : The harmful effects caused by the presence of detergents in water are:
  - ◆ formation of foam, which hinders natural or artificial purification, concentrates impurities and is liable to spread bacteria and viruses; anionic detergent concentrations of 0.3 mg.l-1 and over are sufficient to produce a Stable foam,
  - ◆ formation of a barrier film on the surface, which slows down the transfer and dissolution of oxygen in the water, even when there is no foam,
  - ◆ a soapy taste, at concentrations well above the foam point,
  - ◆ higher phosphate content due to the combination of polyphosphates with surfactants, leading to the eutrophication of lakes and the growth of plankton in rivers; in some countries, a large proportion of polyphosphates is replaced by NTA (nitrilotriacetic acid),
  - ◆ a gradual increase in the boron content of surface and groundwater supplies, due to the large quantities of sodium perborate used in detergents.

Detergents do not kill bacteria, algae, fish and other forms of river life, so long as the concentration does not exceed 3 mgl^{-1}.

Lastly, the enzymes recently added to detergents have no ill effects either on the receiving water or at purification plants. The introduction of detergents which are at least 80% biodegradable has led to a very marked improvement, at least in the case of anionic detergents, which can be successfully checked.

Non-ionic detergents raise problems because they stimulate the formation of foam by anionic detergents and then stabilize it. The non-ionic products now used are resistant to biodegradation, particularly in cold weather.

RADIOACTIVITY

The ingestion of radioactive products can have a somatic effect on man, causing malignant tumors, or mutations that might appear in future generations.

The water produced must meet regulations in each country. Treatment of water includes precipitation by lime, ion exchange or reverse osmosis.

INDUSTRIAL EFFLUENTS

Types of Effluents

Whereas the nature of domestic wastewater is relatively constant, the extreme diversity of industrial effluents calls for an individual investigation for each type of industry and often entails the use of specific treatment processes.

Therefore, a thorough understanding of the production processes and of the system organization is fundamental.

Origin of the Effluents

There are four types of industrial effluents to be considered:

General Manufacturing Effluents

Most processes give rise to polluting effluents resulting from the contact of water with gases, liquids or solids.

The effluents are either continuous or intermittent. They even might only be produced several months a year (campaigns in the agrifood industry, two months for beet sugar production, for example).

Usually if production is regular, pollution flows are known. However, for industries working in specific campaigns (synthetic chemistry, pharmaceutical and parachemical industries), it is more difficult to analyse the effluents as they are always changing.

Specific Effluents

Some effluents are likely to be separated either for specific treatment after which they are recovered, or to be kept in a storage tank ready to be reinjected at a weighted flow rare into the treatment line.

Such is the case in:

- pickling and electroplating baths; spent caustic soda; ammonia liquor from coking planes,
- condensates from paper production, mother liquors from the agrifood industry,
- toxic and concentrated effluents.

General Service Effluents

- Wastewater (canteens, etc.).
- Water used for hearing (boiler blow-down; spent resin regenerants).
- Sludge from the treatment of make-up water.
- Refrigerating water blowdown.

Intermittent Effluents

These must not be forgotten; they may occur:

- from accidental leaks of produces during handling or storage,
- from floor wash water,
- from polluted water, of which storm water may also give rise to a hydraulic overload.

Characterization of Effluents

For the correct design of an industrial effluent treatment plant, the following parameters must be carefully established:

- types of production, capacities and cycles, raw materials used,
- composition of the make-up water used by the industrial plant,
- possibility of separating effluents and/or recyling them,
- daily volume of effluents per type,
- average and maximum hourly flows (duration and frequency by type),
- average and maximum pollution flow (frequency and duration) per type of waste and for the specific type of pollution coming from the industry under consideration.

It is often helpful to be informed about secondary pollution, even if is rare, since it can seriously disturb the working of certain parts of the treatment facilities (glues, tars, fibres, oils, sands, etc.).

When a new factory is being designed, these parameters will be ascertained after analysis of the manufacturing processes and compared with data from existing factories.

Knowing the composition of the make-up water is often necessary.

Specific Pollution Factors

The principal types of pollutants are set out below, classified in accordance with the types of treatment to which they may be subjected:

Insoluble substances which can be separated physically with or without flocculation

- Floating greasy matter (greases, alipharic hydrocarbons, tars, organic oils, resins, etc.).
- Solids in suspension (sands, oxides, hydroxides, pigments, colloidal sulphur, latexes, fibres, filtration aids, etc.).

- **Organic substances separable by adsorption**
  - Dyes, detergents, miscellaneous macromolecular compounds, phenolared compounds, nitrated derivatives, chlorinated derivatives.

- **Substances separable by precipitation**
  - Toxic and nontoxic metals, Fe, Cu, Zn, Ni, Al, Hg, Pb, Cr, Cd, Ti, Be, which can be precipitated within a certain pH range, and sulphides.
  - PO_4^{2-}, SO_4^{2-}, S_3^{2-}, F^-.

Substances separable by deaeration or stripping

- H_2S, NH_3, SO_2, phenols, light or aromatic hydrocarbons, chlorinated derivatives.

- **Substances which may require a redox reaction**
  - CN', Cr_{VI}, S^{2-}, Cl_2, NO_2^-.

- **Mineral Acids and bases**
  - Hydrochloric, nitric, sulphuric and hydrofluoric acids.
  - Miscellaneous bases.

- **Substances which can be concentrated by ion exchange or reverse osmosis**
    - Radionuclides such as I*, Mo*, Cs*.
    - Salts of strong acids and bases, ionized organic compounds (ion exchange) or non-ionised organic compounds (reverse osmosis).
- **Biodegradable substances**
    - For example, sugars, proteins, phenols. After acclimatization, some organic compounds such as formaldehyde, aniline, detergents and even aromatic hydrocarbons as well as some mineral compounds ($S_2O_3^{2}$, SO_3^{2-}), may be oxidized.
- **Colour**

Industrial effluents may be heavily coloured. This colouration is due to colloids (pigments, sulphides) or to dissolved substances (organic matter, nitrated derivatives).

From an analytical viewpoint, the following should be pointed our:

(1) The ratio of COD to BOD; in industrial effluents often differs very substantially from that of MWW. It changes during the stages of treatment, the final COD sometimes reaching a value more than ten times that of the corresponding BOD_5.

(2) The presence of very active toxic substances may conceal that of biodegradable substances and thus seriously falsify the measurement of BOD_5.

DISCHARGE STANDARDS

Discharge standards vary greatly:

- pollution factors are more numerous than in MWW and must be determined industry by industry :

- regulations concerning specific factors (hydrocarbons, heavy metals, F^-, CN^-, phenols in particular) rely on different methods of measurement and are enforced to different degrees depending on the country.

Effluents may be discharged directly into the natural receiving waters or into city sewers which convey them to biological facilities where they should not impair normal operation.

If it is now common practice to set limitations on the concentration of effluent, then regulations concerning maximum quantity of effluent produced per day or per unit manufactured are becoming more common. Monthly averages and daily maxima should also be considered.

Tables 2 and 3 demonstrate the complexity of this problem.

Table 2 : Measuring methods and the range of standards possible for some organic compounds.

Compounds encountered	Methods standards ($mg.l^{-1}$)	Range of
Free cyanides (Epsiein)	- T 90.108 (F) - ASTM D 203682 C	0.2 - 1
Total cyanides (except SCN)	- T 90.107 (F) - ASTM D 203682 A	2 - 6
Phenols transportable in steam (DAAP)	- T 90.204 - API 71657	0.5 - 5
Total phenols		5 - 10
Insoluble hydrocarbons (precipitation on floc, extraction, gravimeny)	- T 90.202 (F) - HMSO 1972 (UK) - API 73253	2 - 10
Total hydrocarbons in refinery (spectro-photometry, IR, 2 peaks)	- T 90.203 (F) - CONCAWE I/72 - S.M. AWWA 502 B	15 - 30
Anionic surfactants	- ASTM D 2330	2 - 10

Sometimes a certain amount of leeway is foreseen in the application of standards in the case where strict compliance would lead to an economic "impossibility". In any case, the standards must take into account the sensitivity of measuring methods and the technical possibilities of treatment.

Table 3 : Examples of standards relating to metals ($mg.l^{-1}$). These norms often apply to total metal, sometimes ro ionized metal.

Total metals	FRANCE Electropla-ting surface treatment	NETHERLANDS Garbage incineration (GSE) (GSE)	F.R.G. Sulphur removal from smoke	SWITZERLAND Discharge into lakes
Ag	.	0.1	-	0.1
Al	5	-	-	10
Cd	0.2	0.05	0.05	0.1
Cr_{III}	3	0.2	0.5	2
Cr_{VI}	0.1	-	-	0.1
Cu	2	-	0.5	0,5
Fe	5	-	-	2
Hg	-	0.01	0.05	0.01
Ni	5	0.5	0.5	-
Pb	1	1	0.1	0.5
Se	-	-	-	2
Zn	5	0.5	1	2
Vd	-	-	-	-

EFFECT OF WATER RECIRCULATION AND CLEAN TECHNIQUES

Their development in two successive stages has, in many cases, led to a reduction in the volume of effluents and in the pollution flow.

Recirculation

Recirculation, first used in cooling units to reduce the amount of water used, was subsequently used to control effluents. It has been used very extensively in the iron and steel industry (whose previous consumption of 200m^3 per tonne of steel has now been reduced to 5 m^3 per tonne, if not 3), in paper production (reduction from more than 50-100 m^3 per tonne to less than 5), and in the manufacture of fibre-board.

This is feasible as long as the main pollution is insoluble and can be removed by a simple physical-chemical procedure either in the system or on a by-pass.

Clean Techniques

Pollute less by producing more efficiently is the environmentalist's challenge to industry which must be taken up in order to optimize manufacturing processes.

The many measures undertaken and their successes are known; only the most significant examples involving just liquid effluents are mentioned.

The replacement of a gaseous emission by a liquid effluent thanks to scrubbing must not entail a transfer of pollution and should result in a simpler way of purifying the effluent or in a direct recirculation of the wash waters into the process line (ammonium nitrate factories for example). This may be carried out by several means:

- **Eliminating the effluents** by developing new "dry" procedures such as may be the case in surface treatment:
    - replacing chromium baths by hot ionic nitriding,
    - replacing cadmium baths by the application of aluminium in ionized steam form,
    - replacing galvanization by Rilsanising (plastic).

- **Separation and possible recovery** of dissolved raw materials which are toxic or costly. For example:
  - **solvents separated by distillation:** the manufacture of paint (HC), of sulphonated resins (dichloroethane), pharmaceuticals (ethanol), tawing of hides (petroleum),
  - **chromium bound to resins:** stabilisation of chromium baths by fixing of Cr^{3+} ions on resins, double ion exchange on rinse water after passivation with recycling of the eluates in the baths.
- **Separation of suspended compounds** from manufacturing processes, and reintegration, whenever possible, into the process:
  - sludge after settling in cardboard production,
  - oils from foodstuffs and margarine refineries,
  - greases and proteins from slaughterhouses.
- **Separation of dissolved compounds** synthesised during processes:
  - phenol from spent soda with liquid—liquid extraction by a recycled oil cut,
  - ammonium from formation water in coking plants or from amino acid production, separated through steam stripping and later recovered through condensation or sulphadon.

AGRIFOOD INDUSTRIES (A.F.I.)

The characteristics common to all effluents from foodstuff industries are essentially organic and biodegradable pollution, and a general tendency to rapid acidification and fermentation. All these effluents are treated primarily by biological methods, but the medium often becomes deficient in nitrogen and phosphorus.

Piggeries

The amount and degree of pollution depend on the methods of stock breeding, the method of cleaning the sties, the time spent in the sties and the type of feed used.

Discharge per pig per day	Cleaned hydraulically	Cleaned dry
Water	17-251	11-131
BOD_5	100 - 200	80 - 120 g
TKN	18 - 35 g	
COD	300-500g	

In hot countries, spraying of the pigs raises the above volume of water.

Manure analysis in $g.l^{-1}$

SS	30 to 80
COD	25 to 60
BOD_5	10 to 30
TKN	2 to 5
Total NH_4	3 to 4
Cl	0.8
SO_4	1.5 to 2
Malk.	1400 to 1500 Fr. deg.
pH	7 to 8

DETRITUS REMOVAL

Simple Manual Screens and Traps

The simplest form of screen needs consist of nothing more than a wire mesh basket through which effluent can flow, the mesh size being dictated by the nature of the material being handled. A plating shop, for instance, might expect to capture nuts and bolts, metal swarf and pieces of plastic from jigs, locating the traps at the outlet of each plating tank to prevent the need to drain rinses and dragouts for a vital piece of lost work.

Even for similar effluents which contain low suspended solids, a mesh trap somewhere in the drainage system is essential, and should be given occasional but regular inspection. Workers in production areas are just as likely to lose rings, loose change, watches, hairclips and even wallets as those in the offices!

Food and dairy processing often produces fat and grease particles which can block pumps and produce odour problems in tanks. An effective wire mesh screen or series of baskets at the inlet to treatment plant will remove this, but such items need regular inspection and cleaning as they often rapidly become blocked by sticky particles, and a mechanically raked fine screen may be more suitable here.

A simple screen consisting of a disposable sack made with 3mm terylene mesh is available from Copa Products Ltd, one of several suppliers; it has been widely used on smaller sewage treatment plants to prevent the discharge of plastics, rag and grease balls from storm tank overflows, and in the dosing chambers of percolating filters, where pricking out the filter arms to remove such debris is a regular maintenance task. The sacks are attached to weirs on tanks or in the dosing syphon chamber, and have a self-cleaning action, whilst also breaking down faecal matter. A version for marine discharges is supplied with an agitation system to assist the solids break-up process. They could be equally effectively incorporated in a gully on the factory floor, or at the overflow of drag-outs and settlement tanks.

A recent innovation at sewage treatment plant is to skim the surface of secondary settlement tanks with submersible pump sets to capture and disperse scum. The device can be modified to remove plastic that has passed unscathed through activated sludge treatment, collecting the material in a mesh sack suspended over the stilling box.

This is a successful, if 'last ditch', method of preventing these materials entering the watercourse. This arrangement, a final settlement tank and the volume collected in less than 24 hours.

Industrial production areas generating gross solids must be fitted with floor gullies covered by gratings with 12—20mm square slots which are screwed or bolted down, and certainly not easily lifted. This removes the temptation to hose everything on the floor down the drain at the end of a shift or working day — or even worse, to clear a temporary flood by lifting the gratings.

There is no finer way of blocking the drainage system of the factory than by hosing down powdery or granular material that then coagulates and sets overnight in pipe bends and dead ends. The sewerage and sewage operations sections of the water companies also take a dim view of lumps of concrete, meat hooks, bone, plastic swarf and pieces of wood in the public foul sewer; ultimately these can jam pumps or screens with expensive results, and may cause flooding for which the water companies become liable if it causes pollution. They therefore tend to devote some effort to recovering costs from the original owners of the material and to prosecuting guilty parties for any pollution indirectly caused.

A simple and very fine rundown screen (0.25-0.75mm aperture) (one such is manufactured by Environmental Engineering Ltd.) is useful for continuously screening solids from agricultural waste containing straw, vegetable washings, peelings and similar food processing waste, and textile washings containing lint, fibres and very fine grit. This device has no moving parts, solids being continuously collected and often quite dry.

Some headloss is inevitable with any screening device, and in the factory environment, the placing of screens in the flow should be checked carefully to prevent any backing-up or flooding. Draining effluent to a sump, followed by pumping to a higher level for treatment is a solution demanding highly reliable pumps, and is best considered in combination with other treatment apart from simple screening, in view of the energy costs.

Mechanical Screens

There has been considerable advancement in the range and performance of mechanical screens in the last decade, giving the designer a wide choice to suit a number of environments. Development has been driven by both economical and environmental considerations, and treatment process changes. Sewage works have been progressively automated—entirely demand on some of the smaller plants — and now rely on the travelling gang for periodic attention and emergency call-out via a telemetry link. Some of the older preliminary treatment equipment at the works inlet requires near-constant attention, particularly during storm conditions when a lot of debris is flushed from the sewers, With few exceptions, screens are driven by a motor/gearbox combination, typically 2hp which is often enclosed within the body of the screen. Health and Safety requirements dictate enclosure of moving parts.

A number of factors need consideration when choosing screens. Choice of type and mesh size will be dictated by a number of on-site and location factors. It is essential that a reasonably comprehensive survey of the size range of detritus is carried out, as sewage often has a unique local quality such as a preponderance of rag, excessive fat or farm materials like straw. Some of the more recent sewage treatment package plants need no primary sedimentation tanks and are of sufficient capacity to avoid the need for flow balancing, but these do benefit from fine screening and grit removal at the inlet.

Extended aeration in oxidation ditches, for example, uses a fairly slow circulation rate of mixed liquor in which grit will readily settle and rag bind into lengths of rope if not previously removed at the works inlet. Reductions in aeration tank capacity of 25% have been experienced after five years' "operation of these systems at sites connected to uncombined sewerage systems, *i.e.* from which surface water is apparently excluded. Effective fine screening and grit removal are thus particularly important for extended aeration systems.

Breakthrough of plastics from the older types of coarse bar screen (more than 50mm mesh) poses particular problems, and has necessitated the fitment of wire baskets or surface skimming pumps, as already mentioned, to final tanks and effluent chambers on many works to comply with the prohibition on plastics being discharged to the aquatic environment.

It is obvious that coarse bar screens (more than 50mm mesh) are no longer appropriate choices for marine discharges and neither is any form of screenings maceration which returns the material to the flow, albeit as smaller particles. Screens of nominal mesh size 15-18 mm only retain about 50% of the screenings load and are equally ineffective in marine applications, where much finer screening (to 6mm or less) is now required so that the resultant liquid discharge contains no readily identifiable solids.

Most marine discharges are remote and sites must now contain adequate storage facilities for collected screenings and grit, and be completely automated. Thomas et al. (1989) provide a useful assessment of a number of screen types that could be used at sea outfalls.

The practice of macerating screenings and returning them to the flow has declined recently in favour of dewatering and bagging, with, at some installations, on-site incineration. Incineration of a material containing a lot of plastics places stringent emission and combustion control requirements on the process and may be uneconomic on a small scale.

Screenings incineration plant is the subject of control under the Environmental Protection Act 1990, Part 1.

High reliability is expected of mechanical screens and one practical consideration is their elevated and exposed position on many works inlets. Freezing up in cold weather is a frequent operation problem, encouraged by the damp atmosphere directly above channels. Some form of thermostatically controlled frost protection including trace heating and insulating washwater supplies, drive chains, gearboxes, hoppers and compactors and condensation heaters in motors should be considered essential in the specification for sites likely to experience frost or freezing fog.

Headloss through coarse screens is typically 50mm, and may be as much as 150mm across a fine 4mm screen, similar to the comminutor that was popular in the 1970s. Choice may therefore be influenced by slight gradients in approach channels and a shallow slope to the works site.

The coarse bar screen is an essential component of the inlet works at many larger sewage works connected to a sewerage system where the sewers are more than 300mm in diameter, combined sewerage systems and those with high infiltration. The primary function of this type of screen is to protect pumps and other equipment in subsequent processes. Some of the largest works are frequent recipients of items of clothing, timber and metalwork like bicycle frames brought down by storm flows. The bars are typically 20-50mm apart, and collect rag, large debris and fat balls quite effectively, trapping smaller plastics floating at right angles as the screen blocks.

Operation is automatic, and usually initiated by the differential head that builds up either side of the bars as they become occluded. This is detected by float switches or ultrasonic liquid level detectors, and most systems can be adjusted within a 25-75mm head difference range. Timed screen operation is also possible, but not appropriate if flows and loads vary widely, as they do at most sewage treatment works.

Most manufacturers can supply bar screens for a standard range of channel widths, and different bars spacings are available, down to 12mm. Other variations on the same theme include stepped bar screens some of which are continuously cleaned by a polypropylene brush which will effectively reduce the amount of fat clinging to the bars. The drive consists of a motor of about 2hp, gearbox(es) and control equipment linked to level detection. The drive is often a chain on enclosed screens or external oscillating arms and must contain a shear pin.

Fine Screens

Most fine screens are cylindrical, being slowly rotated by an electrical drive similar to bar screens. Diameters vary, material is usually of stainless steel, and the whole structure should be of sufficient strength to withstand operators using manual methods to remove colloidal matter. Fine screens can be used to relieve overloaded primary sedimentation tanks by removing some of the potentially settleable solids, and may reduce the biological load to the plant, but depending on the size range of debris arriving on site may blind very quickly. Again, a site survey of screenings is essential before making a decision.

Fine screening has been used in the food industry for some time; and it has become widely used in sewage screening recently to remove plastics and reduce particle size. Mesh size covers the range 0.25-15mm. An aperture size of 6mm has been adopted as the maximum acceptable for Consent applications dealing with marine discharges designed to protect bathing use.

When used to screen sewage, fine screens capture large quantities of debris with a high faecal content; if they are replacing coarser bar screens, this must be allowed for by extra storage capacity and washing/dewatering facilities. An installation at a sewage works where there is no primary sedimentation, the main treatment system being an oxidation ditch.

Most fine screen designs are variations on a theme. The drum screen admits flow from the outside, trapping debris on the outside surface to be scraped or pressure-washed off. Effluent discharges axially from inside the drum. Cup screens work in the reverse manner and solids are trapped on the inside drum surface to be washed off into a channel inside the drum. The microstrainer, used as a tertiary treatment system on sewage works and an initial coarse screen of reservoir water for potable supply is an example. It is essential that the screen is continuously cleaned.

Other variations contain nylon brushes and polypropylene scrapers, and are particularly appropriate for fat-bound materials, *e.g.* pet foods and abattoir waste where screen blinding occurs. A 40% reduction in suspended solids (SS) and a 30% reduction in BOD is claimed by Lockertex Ltd., one of the manufacturers of drum screens.

Mesh blinding is a frequent problem with fine screens; the apertures may become completely blocked by fine strands of solids weaving in and out of them. This is termed 'hair-pinning'. Sewage containing a high rag and hair content is a typical culprit. Doubling the drum rotation speed, significantly increasing wash water pressure and mechanical brushing have all been used with success to combat mesh blinding. Substantially increasing the screen mesh thickness also makes a significant improvement. Some claims have been made that plastic mesh is less prone to the problem than metal. The theoretical 'design aspects of cup screens have been well reviewed by, for example, Stamper and Graham (1981).

Screenings and Detritus Disposal

Screenings collected by any of the above methods are gathered into a trough or screw conveyor flushed with sewage or works effluent, or a hopper from which excess liquid can drain, and are further treated by either maceration, dewatering and compaction, or incineration.

Maceration

In this method the material is passed through an in-line disintegrator, which reduces the size of the detritus, and then returned to the flow. This was a popular technique in the 1960s and 1970s; the theory was that reduced material would settle better in subsequent primary sedimentation and cause less damage en route. The presence of increasing amounts of plastic in sewage defeats the object, and the method is not suitable for any secondary treatment system (for example, extended aeration) without primary sedimentation or grit separation. The macerator unit itself is a high-maintenance item and can be readily damaged or jammed by bits of wood and rag balls.

A contemporary version consists of a macipump discharging screenings from a collection hopper into a centrifugal dewaterer. A very consistent product is obtained, the device dealing well with plastics. It is often teamed with a fine drum screen to collect the screenings initially, with grids to prevent the passage of any large items (over 50 cm). Figure 3.2 outlines the installation of four units at a sewage works with a dry weather flow of 12000 m^3/day, and reducing the material collected by two fine 6 mm screens. The product discharged to a skip below the macipumps and dewaterer is less than 15% organic matter. Suppliers of these units include Haigh Engineering Co. Ltd.

Dewatering and Compaction

This is a favoured method, as it reduces the volume for eventual disposal. Screenings are moved by a screw auger to a hopper, loaded into a second conveyor and either compacted by a ram or driven up an inclined plane by auger, the water draining from the lower end. The discharge is often continuous and into a trailer. A volume reduction of 75% is possible with a compactor. A number of manufacturers supply such equipment, often as a complete package with a screen set. In this state, screenings are a potential odour problem and quickly start to decompose. On many small sewage works, screenings are buried on-site and quickly rot down.

The author saw an interesting experiment in the 1970s: screenings were bagged in black airtight sacks; on inspection after a year they had completely composted to a fine dark material, and the majority of the plastics either powdered or disintegrated.

Handling larger volumes presents a disposal problem. Bagging and landfilling has a transport cost and available sites that will accept what is defined as a hazardous waste may be limited. Nevertheless, this is a common disposal route, and co-disposal is usually the cheaper off-site option. On remote sites, some form of bagging is essential if vermin are not to be attracted coupled with a regular removal routine.

Several practicalities arise with bagging. The plastic bags used should be strong enough to withstand manhandling without tearing, and small enough to lift and load by hand (25kg). A few drainage holes in the sack base will allow any further water to drain away, as after a week of storage, some natural dewatering takes place. The bagging unit needs partial enclosure or shelter, as bags are prone to freezing in northern climates due to the exposed areas of the conveyor and compactor and the elevated aspect of many works inlets.

Incineration

This ultimate disposal route is attractive because if sufficiently dewatered to more than about 35% dry matter, screenings will burn autothermically. Some practical experiences are presented by Pace and Price (1982).

Stringent emission standards have been placed on incinerators in Part 1 of the Environmental Protection Act (1990). Contemporary screenings containing a lot of plastic will require incineration at temperatures above 1000°C to prevent dioxin formation', these will generate hydrogen chloride gas and bring about the need for flue gas scrubbing.

Incineration costs are therefore high compared to those of landfill, but if carefully engineered incineration does provide on-site disposal and the prospect of heat recovery. The low ash volume generated would be acceptable for landfilling, but some leachate trials for metals, particularly zinc and cadmium, would be advisable.

INDUCED POLLUTION

Introducing a reagent into water can lead to two types of pollution: impurities from the reagent itself, and products produced by a reaction of the reagent with organic matter in the water.

Impurities due to the Reagent

In many countries, the use of a reagent must be approved by the health authorities. Regulations may determine, for each reagent, the maximum concentration of impurities allowed. A precise analysis of the products must be undertaken. In the event that the presence of impurities is established, it is important to make sure that the planned chain of treatment will remove them.

- **Inorganic Coagulants**

Some coagulants are made from minerals in which the level of impurities is noteworthy: acid attack on bauxite, making aluminium sulphate; attack on metal structures, making ferric chloride. This attack also dissolves impurities (tungsten, manganese, arsenic).

Other coagulants are made from by-products coming from another industry. Chlorinated copperas made from ferrous sulphate which comes from the manufacture of titanium, may contain a substantial amount of manganese.

- **Polyelectrolytes, Coagulant Aids**

Synthetic polyelectrolytes are made by polymerization of monomers (polyacrylamides, polyamines, etc.). In drinking water treatment, each country's regulations may establish the type of monomer that may be used, the maximum monomer content in the polymer and the maximum treatment rate allowable (0.5% of acrylamide monomer in the case of polyacrylamides, for example).

- **pH Correction**

The impurities in the lime and caustic soda must be analysed to make sure no mercury is present (caustic soda from a membrane procedure).

- **Disinfection using Chlorine**

Since disinfection by chlorine is the last stage in the treatment process, it is important to verify the purity of the product before disinfection.

- **Stripping Air**

The air used in certain stages of the treatment may introduce undesirable elements: impurities in the atmosphere, exhaust gas, fumes, bacteria, etc.

Impurities due to a Reaction between the Reagent and Organic Matter in Water

During the oxidation phases which occur all along the treatment line, oxidants (ozone, chlorine dioxide, chlorine) can react on organic matter present in the water. Chlorine, especially, reacts with certain of these (precursors) leading to the appearance of haloforms.

This reaction can occur directly by means of the chlorine in the water which is in the form of ClO^-, or by means of an- other halogen (bromine or iodine) which could have been replaced by chlorine (resulting in XO^-, which would then react with the organic matter to form organo-halogen compounds).

The basic reaction to form the trihalomethane compounds is the following:

$$R{-}\overset{\overset{\displaystyle O}{||}}{C}{-}CH_3 + 3OX^- \rightarrow R{-}\overset{\overset{\displaystyle O}{||}}{C}{-}O^- + CHX_3 + 2OH^-$$

Carbonaceous organic substances capable of causing this reaction are, for the most part, methyl ketones, or, to put it more generally, all organic products which by means of oxidation can be oxidized into methyl ketones.

The chlorination of organic compounds also leads to the formation of other compounds, some of which, until today, remain unidentified.

- **Hazards**

All these products are suspected of being carcinogenic. It is important that the plant and its operation be such as to produce water that contains as few trihalomethanes and organochlorinated compounds as possible.

It must be remembered that the attempt to reduce the level of organo-chlorinated compounds must not compromise the thorough disinfection of water for distribution.

POLLUTION FROM EXTERNAL SOURCES

Substances which make up several of the products used for water treatment can be a source of pollution. As an example of this, rake the case of ion exchange resins or that of membranes. Certain regulations have established a list of products authorized for use in the manufacture of these resins (or membranes); rests have been worked out to ascertain that there is no release while the product is being used in the treatment line (removal of nitrates by ion exchange resins, for example).

Materials making up the supply system may also be a source of pollution: solvents used for the painting and coating of structures, monomers and additives used in plastic pipe and tube manufacture, metals used in valves and fittings or even lead used in the manufacture of pipes and tubes (which might cause lead poisoning), especially in home plumbing systems. Some of these materials may release organic carbon which may serve as food for common bacteria, thus aiding their proliferation in the supply system; they may also cause the water to taste bad.

MUNICIPAL WASTEWATER

Source and Collection Method

Municipal wastewater is made up of:

- domestic wastewater or sewage,
- rainwater.

Municipal wastewater is mainly domestic (sewage, etc.); the industrial portion generally keeps pace with population growth.

Factories which discharge an unproportionally large amount of pollution, or pollution requiring special treatment, usually have their own purification system. Depending upon the level of this treatment, the industrial effluent will join the receiving medium or the collection system and treatment system.

The daily volume and pollution of the industrial effluents are frequently expressed in population equivalents and correspond to the per capita amounts of domestic wastewater.

Sewerage comprises collection, transport and treatment of urban effluents. The main two collection systems used in sewerage are of the continuous operation type:

- the combined system conveys domestic wastewater and rainwater in the same line;
- the separate system is made up of two different pipe networks: one for rainwater, which is about the same size as the combined system, and the other, smaller in size, for the wastewater per se.

The first sewerage systems were combined; the separate system is a more recent development. The collection systems found in densely populated areas are seldom all of one type.

The partially-separate system refers to a separate system in which water from roofs and backyards is conveyed to the wastewater network.

The combined systems are equipped with storm overflows that discharge some of the flow directly into a river during rainy periods. In this case, only part of the pollution flow reaches the purification plant. This discharge of pollution into a river during rainy periods constitutes, along with poor self-cleansing of the lines during dry periods, two of the main dis-advantages of combined systems. However their construction is easier to control. An essential property of a network is that it is watertight and keeps our parasites from groundwater or from surface run-off.

Effluents are conveyed, as far as possible, by gravity, however, the topography of a terrain often means that intermediate pumping stations must be installed. These can then be followed by mains of considerable length under pressure. In special cases, the effluents may be collected at the point of emission under pressure or even under vacuum; these methods, developed for small installations, allow pipes to be laid at a higher level.

There are two types of sewerage that do not involve the collective network of sewers:

- individual or grouped sewerage serving one residence or several residences situated very close together, often called self-directing sewerage. The domestic wastewater is conveyed directly to a water purification facility, "flowing" normally into the ground.

- the treatment of night soil in a centralized plant. This method of sewerage consists of storing the most concentrated effluents from each house in watertight pits which are regularly emptied into tank trucks and conveyed to a central treatment plant. This method is still used in some areas where the houses are very old and/or the ground relief such that a wastewater collection network is not feasible. It is sometimes used ro drain off septic ranks. In Asia, this type of collection and treatment of night soil is still in use. The delay in constructing sewage systems is often due to the great density of the population.

It must be mentioned, also, that some countries in Asia follow the ancestral custom of spreading fecal matter directly on crops. Thus, wastewater collected in a sewerage system would have a relatively weak concentration in spire of a low water demand.

Waste Water

The per capita volume of wastewater discharged per day usually increases with the population as a result of changes in life style and a larger participation by secondary and tertiary sectors. The volume also varies according to the area of the world and its level of development. The method of determining the consumption rate of drinking water may also have a bearing on it.

Although, it accounts for less than 1501/person per day in many rural areas, the amount of municipal wastewater produced during dry periods today in France is about 2001/ person per day in towns with populations in the several tens of thousands. The volume is over 3001/ person per day in Paris where large amounts of water are used to wash the streets. In many American or Swiss cities, the volume rises well above 4001/person per day. Frequent catchment in combined sewerage systems of rainwater collected in small screams inflates these ratios. Infiltration water, which very often enters the system unchecked, has the same effect.

These variations can be costly, for the sizing of a purification facility depends very much on its hydraulic capacity.

The amount of municipal wastewater varies throughout the day. In less densely populated towns there are two flow peaks; in more heavily populated towns there is only one. Sociological development is causing the latter type to become more generalized (see Figure 2.4). The shorter the sewerage system and the smaller the population served, the greater the size of the peak. The presence of numerous pumping stations on a system may have a comparable effect.

In many countries, an average diurnal flow during dry weather may be represented by Qd. If Qj is the daily flow, Qd lies most often between:

Qj/14 and Qj/18

The peak dry weather flow Qp in a separate system can be calculated by the formula:

$$QP = Qm\left(1.5 + \frac{2.5}{\sqrt{Qm}}\right) \text{ in } l/s^{-1}$$

with Qm = daily average = $\frac{Qj(\text{in } l)}{86{,}400}$

and up to a peak coefficient

$Qp/Qm \leq 3$

Besides the daily variations, substantial seasonal variations in wastewater flow are becoming more frequent, even in large cities. These variations are closely tied to vacation periods during which there is a large exodus of city dwellers heading for vacation spots. In .very developed countries, many city dwellers leave their principal residence during the winter sports season as well as in summer; in this case, they use three different sewerage systems each year. This is not without consequence on the demand for collective facilities.

Rainwater

The rainwater flow depends:

- on the incline and on the surface of the drainage area,
- on its degree of imperviousness,
- on the rainfall.

Various formulas have been put forth, one of which is Caquot's.

For European urban areas of several hundred hectares, the maximum amount of run-off water over a five-year period is about 200 l/s.ha.

ASSESSMENT OF POLLUTION

Suspended solids loading contributed by raw sewage, per head per day, is generally estimated as follows:

- **separate system:** 60 to 80 g, including 70% volatile solids,
- **combined system:** 70 to 90 g, including 65% volatile solids.

The per capita quantity of pollution increases as the standard of living rises and usually as the population grows. However, the sewage flow increases faster than the loading, so that the sewage tends to become less concentrated. If the sewerage system is badly constructed, infiltration water can get in and aggravate this tendency.

The figures cited above are for SS after screening and grit removal and do not include material collected during preliminary treatment, which is roughly as follows:

- **Screening:**
  - ◆ 2 to 5 dm^3 of screenings per person per year in the case of screens with bars approximately 40 mm apart,
  - ◆ 5 to 10 dm^3 for screens with bars approximately 20 mm apart,
  - ◆ 10 to 15 dm^3 for screens with bars approximately 6 mm apart.

The water content of these screenings is about 70 to 80% after natural drainage and pressing.

- **Grit Removal:**

In a combined system, the volume of grit per head per year is about 12 dm^3 for an area of detached houses, and 5 dm^3 for areas with high-rise apartment buildings.

- **Rainwater**

It is interesting to make an assessment of the pollution conveyed by rainwater alone. It is difficult to establish an overall rule since there have not been many in-depth investigations into the matter and on site conditions play a large role. The pollution varies considerably during the course of the precipitation; it is usually more concentrated at the beginning of a storm. The matter entrained in the rain, usually being more mineral in nature than that found in sewage, does, however, contain hydrocarbons; lead, from motor fuel, is also frequently present.

A study made in France by the Service Technique de l' Urbanism (S.T.U.), assesses the yearly pollution flow at 300 to 1100kg SS per hectare of drainage area and from 800 to 2650 kg SS per hectare of watertight drainage area.

Carbonaceous Organic Pollution

The BOD loadings contributed by sewage, per head per day, can be estimated, after preliminary treatment, as follows:

◆ separate system: 50 to 70 g,

◆ combined system: 60 ro 80 g.

Roughly one third of this pollution is dissolved; the remaining two thirds are tied to particles (settleable or not). In combined systems the percentage of organic pollution that is settleable is usually higher than that found in separate systems.

The COD/BOD_5 ratio in sewage is usually between 2 and 2.5.

The TOC, due to the nature of the analytical method, cannot identify paniculate components.

- **Rainwater**

The aforementioned S.T.U. study proposes the following assessment of annual pollutant flow in rainwater:

	Drainage area	Watertight drainage area
BOD kg per head	30-85	80-210
COD kg per head	160-460	470-1,100

Other Components

- **Nitrogen**

The total nitrogen content (TKN) amounts to about 15 to 20%. of the BOD; in domestic effluents. The daily contribution of TKN is between 10 and 15 g per head.

- **Phosphorus**

The daily contribution of phosphorus is about 4 g per head. It is essentially due to human metabolism and the use of detergents. Moreover, the amount varies according to the day of the week.

- **Surfactants**

Widespread use of biodegradable detergents has lessened the operating problems plants have had with foam; but their use continues to grow.

- **Micro-elements**

The most harmful elements are heavy metals. Their main source is industry. Copper, zinc, cadmium, chromium, lead, mercury and nickel are the pollutants most frequently encountered. The level of these elements is usually less than $1 mg.l^{-1}$. Discharging toxic products into sewers and especially into natural receiving waters is prohibited by law (*e.g.*, cyanides, cyclic hydroxyl compounds).

Table 4 gives the value ranges for average characteristics of MWW in France:

Table 4 : Characteristics of French MWW.

Parameters	Variation range	Settleable fraction
pH	7.5 to 8.5	
Dry solids $mg.l^{-1}$	1,000 to 2,000	10%
Total SS $mg.l^{-1}$	150 to 500	50 to 60%
BOD, $mg.l^{-1}$	100 to 400	20 to 30%
COD $mg.l^{-1}$	300 to 1,000	20 to 30%
TOC $mg.l^{-1}$	100 to 300	
TKN $mg.l^{-1}$	30 to 100	10%
$N\text{-}NH_4^+$ $mg.l^{-1}$	20 to 80	0%
$N\text{-}NO_2^-$ $mg.l^{-1}$	< 1	0%
$N\text{-}NO_3^-$ $mg.l^{-1}$	< 1	0%
Detergents $mg.l^{-1}$	6 to 13	0%
P $mg.l^{-1}$	10 to 25	10%

Pathogenic Agents

MWW conveys many microorganisms, some of which are pathogenic: bacteria, viruses, protozoa, helminths. The pathogenic bacteria most often encountered are salmonella, the level of which is somewhere between 10^2 and 10^3 per 100 ml.

The process of identifying pathogenic organisms in water is a lengthy one; indicator germs are used for routine counting. The most common indicator germs used are the total coliforms and the fecal coliforms (primarily Escherichia coli) and sometimes also the fecal streptococci. In 100 ml of domestic sewage, between 10^7 and 10^8 total coliforms and between 10^6 and 10^7 fecal coliforms have been counted most often.

The fact that coliforms are not found in an effluent does not mean that it does not contain any pathogenic germs. In MWW, indicator germs are used, above all, to estimate the level of decontamination (or of partial disinfection); this is expressed by the number of logarithmic units (power of 10) which a treatment has been able to reduce from the initial count in the raw water.

Oxidation-Reduction Potential - Septicity

Domestic wastewater that is sufficiently fresh has an oxidation-reduction potential of about 100 mV, corresponding, for a pH of about 7, to an pH of about 17 to 21. A potential of +40 mV (let pH = 15 at pH 7) or a negative potential indicates a reducing medium (septic wastewater, putrid fermentation, presence of chemical reducing agents). A potential exceeding 300 mV (pH = 24 at pH 7) reflects an abnormal oxidizing medium. The septicity of the effluent leads to the formation of sulphur (S_2^-) and causes release of H_2S.

Night Soil

The characteristics of night soil may vary greatly, depending on its origin; it comes primarily from two types of watertight pits:

- those with a "water effect" (very small volume of water mixed in with fecal matter), usually emptied once a year; night soil is very concentrated (COD comprised between 12 and 30 $g.l^{-1}$),

- those equipped with a water flushing unit; they are emptied more frequently and night soil is much more diluted (COD about 2 g.l^{-1}).

The available volume of the cesspool space determines the frequency of emptying and does or does not ensure preliminary digestion of the effluents. Table 5 gives the characteristics of these products in several countries.

Pollution Peaks

Usually the pollution level varies at the same rime as the flow, so that variations in the pollution flow are greater than those of the flow alone. This phenomenon is particularly noticeable with SS in combined systems.

Table 5 : Composition of night soil.

Parameters (concentration in g.l^{-1})	Usual variations in France	Usual variations in Japan	Special cases Saudi Arabia (Taif)
pH	7.7 to 8.5	6.4 to 7.9	6.2
COD	2 to 30	8 to 15	1.75
BOD	1.5 to 10	5 to 9	0.42
SS	2 to 10	20 to 35	0.66
Total Kjeldhal	0.5 to 2.5	3.5 to 6	0.17
nitrogen of which ammonia nitrogen	0.4 to 2	3 to 4	0.12

THE PURPOSE OF TREATMENT—QUALITY OBJECTIVE

The purpose of the treatment of effluents before discharge is ro protect the natural receiving waters. Regulations have progressively diversified the tolerable discharge levels to bring them into line with the objectives of river water quality set down by the authorities in charge of water management. The quality objectives rake into account the primary uses of river water in the reaches downstream (fishing, swimming, drinking water).

Keeping in mind the hazards of hydrological variations in rivers, these quality objectives must be respected in the case of every flow that is less than the flow of a given frequency (French method), or during a minimum number of days in the year (English method).

In the case of discharge into estuaries and into the sea, the conditions of discharge that have been adopted take into account the particular activities that take place in that coastal area (bathing, shell-fish farming).

Table 6 : Overall criteria for appraising the quality of river water. (Grid used by the Agence de Bassin Adour-Garonne, France).

	1A	1B	2	3
Conductivity μS/cm at 20°C	≤ 400	400 to 750	750 to 1500	1500 to 3000
Temperature	≤ 20	20 ro 22	22 to 25	25 to 30
PH	6.5 to 8.5	6.5 to 8.5	6 to 9	5.5 to 9.5
SS (mg.l^{-1})	≤ 30	≤ 30	≤ 30	30 to 70
Dissolved O_2 (mg.l^{-1})	> 7	5 to 7	3 to 5	aerobic medium
Dissolved O_2 in saturation %	> 90%	70 to 90	50 to 70	to be permanently maintalned
BOD_5 (mg.l^{-1})	≤ 3	3 to 5	5 to 10	10 to 25
PV (mg.l^{-1})	≤ 3	3 to 5	5 to 8	
COD (mg.l^{-1})	≤ 20	20 to 25	25 to 40	40 to 80
NH_4 (mg.l^{-1})	≤ 0.1	0.1 to 0.5	0.5 to 2	2 to 8
NO_3 (mg.l^{-1})			< 44	44 to 100
Total N (Kjeldahl)	≤ 1	1 to 2	2 to 3	
Fe (mg.l^{-1})	≤ 0.5	0.5 to 1	1 to 1.5	
Mn (mg.l^{-1})	≤ 0.1	0.1 to 0.25	0.25 to 0.5	
F (mg.l^{-1})	≤ 0.7	0.7 to 1.7	0.7 to 1.7	> 1.7
Cu (mg.l^{-1})	≤ 0.02	0.02 to 0.05	0.05 to 1	> 1

Zn ($mg.l^{-1}$)	≤ 0.5	0.5 to 1	1 to 5	> 5
As ($mg.l^{-1}$)	≤ 0.01	≤ 0.01	0.01 to 0.05	> 0.05
Cd ($mg.l^{-1}$)	≤ 0.001	≤ 0.001	≤ 0.001	> 0.001
Cr ($mg.l^{-1}$)	≤ 0.05	≤ 0.05	≤ 0.05	> 0.05
CN ($mg.l^{-1}$)	≤ 0.05	≤ 0.05	≤ 0.05	> 0.05
Pb ($mg.l^{-1}$)	≤ 0.05	≤ 0.05	≤ 0.05	> 0.05
Se ($mg.l^{-1}$)	≤ 0.01	≤ 0.01	≤ 0.01	> 0.01
Hg ($mg.l^{-1}$)	≤ 0.0005	≤ 0.0005	≤ 0.0005	> 0.0005
Phenols ($mg.l^{-1}$)		≤ 0.001	0.001 to 0.05	0.05 to 0.5
Detergents ($mg.l^{-1}$)	≤ 0.2	≤ 0.2	0.2 to 0.5	> 0.5
S.E.C. ($mg.l^{-1}$)	≤ 0.2	0.2 to 0.5	0.5 to 1	> 1
Coliforms (count 100ml)	≤ 50	50 to 5000	5000 to 50000	
Esch. Coli (count 100ml)	≤ 20	20 to 2000	2000 to 20000	
Fee. strep. (count 100ml)	< 20	20 to 1000	1000 to 10000	
Divergence of biotic index from the normal index	1	2 or 3	4 or 5	6 or 7

Notes : (*) Substances excractible with chloroform.

Note on the use of Table 6.

- The quality of water depends on many parameters (temperature, ammonium (NH_4), mercury (Hg), etc.).
- It is usual to judge water quality on the basis of its least favourable parameter.
- This quality is that which, according to the limits in the table, is achieved by at least 10% of the worst samples in this parameter.

Class 1A : This characterises water that is considered free from pollution and is of a quality to meet the most exacting requirements.

Class 1B : Slightly lower in quality, this water may, nonetheless, meet all needs.

Class 2 : "Acceptable" quality: suitable for use in irrigation and industry, it may be used for drinking after extensive treatment. It is generally tolerable for the watering of livestock. Fish may live in it with no ill effects but their reproduction may be impaired. It may be used for water sports as long as there is not excessive contact with the water.

Class 3 : "Mediocre" quality: barely suitable for irrigation, cooling and leisure boating use. This water may support fish culture but it may be hazardous to fish life in periods of low flow or high temperatures, for example.

Unclassifiable: Water which exceeds the maximum tolerable limit in class 3 for one or more values. It is considered unsuitable to most uses and may constitute a danger ro public health and the environment.

THE REUSE OF SEWAGE

Agricultural Use

The use of sewage in agriculture is very old and land disposal was the first purification system. The ground is an efficient filter and one hectare contains up to one or two tonnes of microorganisms. Today the main reason for reusing sewage in cultivation is, more often, to supply water so necessary to arid areas, rather than to purify it in the ground or to contribute nutrients.

Measures must be taken in order to avoid deposits and corrosion in the distribution system; it is advisable, in any case, to have raw sewage undergo preliminary settling. Preliminary biological treatment is also often recommendable. In particular, it substantially reduces the risk of bad odours.

Two categories of risks are associated with the reuse of sewage:

- **Health risks** for close-lying neighbourhoods and for consumers of the produce.

The risks vary greatly depending on the local state of sanitation in the area, farming methods, customs and climate. However, generally speaking, sewage should not be used on or near vegetables that are eaten raw.

The risk is reduced with proper management determining when and when not to spread sewage and the drying of the crop. The use of sewage on hay meadows does not seem to pose any great problem, which is not the case for grazed pasture.

Arboriculture, cereals, beers and oleaginous crops are the types of cultivation most suited. Surface irrigation is preferred to spray irrigation.

- **Risks to the soil and crops** : clogging the soil, increased salinity, introduction of toxins.

Land disposal can alter the physical properties of the soil. In particular, the introduction of an excessive amount of sodium and the absence of leaching (especially in areas of low rainfall) can destroy the soil structure. Knowing the SAR (sodium absorption ratio) of the effluent is, therefore, important:

$$SAR = \frac{(Na^+)}{\sqrt{\frac{(Ca^{2+} + (Mg^{2+})}{2}}}$$

If the SAR nears 10, it signals danger; usually this only happens in certain concentrated effluents (distilleries, sugar mills, cheese factories).

An effluent that has an excessive salinity level ($>2 g.l^{-1}$) also causes trouble and leads to a stricter control on the quantity of water that is spread and the level of salinity as it develops in the crop soil.

The C/N ratio remains essential for crop requirements. The N/P/K usually shows a very excessive amount of nitrogen in domestic sewage. One of the drawbacks to using sewage in agriculture is that it may raise the nitrate level of the ground-water.

Tables 7 and 8 present an example of guidelines for irrigation water adopted in California.

Table 7 : Table of guidelines for the main characteristics of irrigation water.

	Limits to use		
Characteristic	**None**	**Moderate**	**Strict**
Salinity mgl^{-1}	< 450	450-2 000	> 2000
Spedfic conductivity μS.cm^{-1} (EC)	< 700	700-3 000	> 3000
SAR = 0-3	> 700	700-200	< 300
= 3-6	> 1200	1200-300	< 300
= 6-12 EC	> 1900	1900-500	< 500
= 12-20	> 2900	2900-1 300	< 1300
= 20-40	> 5000	5000-2 900	< 2900
Na			
Surface irrigation SAR	< 3	3-9	> 9
Spray irrigation mg.l^{-1}	< 70	> 70	Cl
Surface irrigation mg.l^{-1}			
Spray irrigation mg.l^{-1}	< 140	140-350	> 350
	< 100	> 100	– pH
	6.5 to 8.4		

Wastewater may also be used to irrigate leisure areas such as golf courses and parks, etc. This irrigation is often done by spraying.

The treatment required is therefore extensive and must end in disinfection after the removal of suspended solids and organic pollution.

Table 8 : Table of guidelines for trace elements in irrigation water (1).

Characteristic	Limit value $mg.l^{-1}$	Characteristic	Limit value $mg.l^{-1}$
Al	5	As	0.10
B	0.75	Be	0.10
Cd	0.010	Cz	0.10
Co	0.05	Cu	0.2
F	1	Fe	5
Li	2.5	Mn	0.2
Mo	0.01	Ni	0.2
Pb	5	Se	0.02
Va	0.10	Zn	2.0

(1) These values pertain to land under continuous cultivation with a yearly amount of water of about 1.20 m. For shorter cultivation periods, these values may be raised.

Use in Industry

Following treatment, urban wastewater may be a source of water that is completely suitable for industrial needs, especially for cooling and washing. Such reuse has nowadays a large number of applications.

Very often, exhaustive removal of organic pollution is necessary and biological treatment is then followed by a finishing treatment.

After very thorough tertiary treatment involving, among others, a demineralization phase, wastewater can be used as feedwater for low-pressure boilers. Prolonged studies have shown that this solution is also workable for medium-pressure boilers.

Domestic and Municipal Use

The reuse of treated wastewater in the home or at the city level is possible for various levels of quality and in accordance with a number of working plans:

- partial recycling inside buildings. This use, which has been undertaken in the Far East, involves supplying flushing water for toilets from recirculated wastewater that has been treated,
- supplying municipal systems with wash water (streets, trucks, etc.), and water for fire-fighting. Its usage must not cause impairments to the system (deposits, bacterial proliferation, corrosion, etc.) or cause unacceptable health hazards,
- partial aquifer recharge (filtering beds, etc.),
- setting up underground hydraulic barriers to prevent the intrusion of sea water into coastal aquifers,
- reinjection into the drinking water system. This usage requires a very complete chain of treatment but does not arrive at a technical impasse. It is being used in deserts and has been under study for use in spaceships. Because of the salinity level in the reused water, it is often necessary to mix it with fresh water to meet drinking water standards. If this is not possible, it must undergo desalination. Great care must be taken to rêmove organic pollution, ammonium and bacteriological pollution.

GRIT REMOVAL

Grit removal is the second essential preliminary stage of sewage treatment, but it is comparatively rare in industrial effluent treatment installations as a separate process, being frequently combined with primary settlement. Grit arises in sewage from the erosion of pipework and joints, and infiltration into leaking sewers through sand and gravel strata.

In larger drainage systems where the surface, storm and foul sewerage systems are combined, large volumes of eroded road surface material and general dirt from vehicles add to the flow, and skip loads can be collected at a large works following heavy rain. Ongoing renovation of old sewerage systems involving the separation of surface and foul water provides welcome improvements and will assist economic use and design of capacity at treatment plant, not only in the appropriate sizing of grit and detritus removal facilities but in coping with a more normal flow variation in wet weather. It is important therefore to size grit removal facilities in relation to the type of sewerage system, and to allow leeway for known infiltration rates and probable increase as the pipework degenerates.

Grit removal is essential at sewage treatment works for two main reasons:

(1) to reduce blockage, wear and erosion of pump bodies, bearings, stators and rotors, and pipework, environments where it will act in the manner of a fluidised grinding paste;

(2) to prevent the reduction in available operating capacity that occurs when grit is deposited in sludge tanks and digesters, and in oxidation ditches which are surface aerated and where the flow velocity around the system is inadequate to keep inorganic material in suspension.

The effectiveness of the grit removal stage directly dictates the frequency with which these tanks and digesters require draining and the grit laboriously and expensively dug or jetted out. A 25% reduction in tank capacity arising from grit build-up is not uncommon in less than 10 years of continuous digester operation, and creates operation problems and the potential for works performance sliding below discharge Consent limits.

As a cause of slow deterioration in works effluent quality, this factor should not be overlooked because grit removal efficiency varies widely with flow and is never 100% for any contemporary equipment. The dead volume of a tank arising from retained grit can be fairly simply measured by comparing actual versus theoretical retention times using a suitable tracer such as sodium chloride.

Prior to grit removal, most waste water treatment plant inlet channels are made deliberately narrow, but most are of sufficient capacity to cope with an order of magnitude change in received water flow rate, in order to maintain velocity and retain all materials in suspension.

Conventionally, grit is removed after detritus has been taken out, and a common method is to reduce the flow rate of incoming waste water to a linear surface velocity of less than 0.3 m/s by widening the incoming channel to form a chamber with inlet and outlet baffles. A scraper, top-driven from an overhead bridge, slowly rotates, sweeping the grit into a collecting hopper.

From here, collected material could be pumped or drained out, but as the hopper is normally flooded and in constant use, the grit is usually removed by pushing it up an inclined channel with an oscillating arm with attached scraper blades. Figure 3.3 shows a cross-section through such a detritor commonly found in the water industry. Materials ranging in size between 0.5mm and 15mm (silt to pea shingle) are effectively removed because, being denser than water, they rapidly fall out of suspension. Periodic drainage of this device is necessary to clear grit collecting at the corners of the grit chamber, and which the scraper will miss.

Two other methods of grit removal are often installed at sewage treatment works. The first method comprises a couple of channels in parallel, V-shaped so that as flow increases, so too does available cross-sectional area. The sloping side walls also assist in grit deposition at the base of the channels, from which the grit is sucked by vacuum pump, and ultimately discharged to a dry hopper. This is a land-consumptive device, and has been superseded by the second method, a vortex design offered by several manufacturers including Tuke & Bell Ltd. and Promech Environmental Ltd.

Flow enters and is swirled round in a cone-shaped structure. Mechanical stirring separates organic and inorganic solids, the latter falling to the base hopper. After washing, the grit is removed by airlift pump or screw. The device is compact: 95% removal of 100 mesh particles is claimed, and particles down to 0.15mm may be retained at normal flows.

The variable grit removal efficiency of these devices at higher flow rates is inevitable and has already been mentioned. Since flow rate is highly variable on many sewage works, grit removal equipment represents a design compromise.

A consensus of water industry experience is that the detritor design shown in Figure 3.3 removes about 85% of grit entering the works; analysis of this material shows it to be 60% - 70% inorganic.

The vortex design is much more dependent on the maintenance of air blowing or stirring rates to keep organic and inorganic particles separated, and can quickly fill up unnoticed. Less grit is then trapped. Nevertheless, large volumes of grit are removed by the devices described, particularly during high flows in winter, when road and concrete surfaces and pipework break up from mechanical ;md frost damage, and silted storm drains and overflows are flushed out. Quantities of sand are collected at works near to the coast, arising from infiltration through sandy strata and washing clothes and towels taken onto beaches.

Although some of the above-mentioned devices wash the grit during extraction, it is rarely more than 70% inorganic when of sewage origin and always contains entrained organic solids and food particles like peas and beans, and scraps of plastic that escaped screening. The moisture content varies widely between 15% and 70%. The material is thus valueless, and on most small sewage works sites it is dumped in a convenient hole. An odour nuisance can be generated in summer if fresh material is not buried. Larger sites must observe the Duty of Care Section 34, Part 1 of the Environmental Protection Act 1990 (that such material must be disposed of in a licensed tip).

Industrial users who may find the vortex type of grit separator useful for trapping aggregates like grit and sand will be able to recycle uncontaminated material, but they must check the device regularly, and in particular the nozzles on air-lift pumps.

OIL AND FUEL SEPARATION

Facilities are not normally provided at sewage works fur the removal of floating oils because they are specifically excluded or, typically, limited to levels of 500mg/l in industrial Discharge Consents to the foul sewer. The arrival of diesel, gas oil or petrol at a works indicates a spillage or accident requiring immediate investigation and possibly flow diversion to storm tanks. Fortunately, this type of pollution is easily spotted and traced to source by the judicious lifting of manhole covers throughout the sewerage system, the material always leaving traces on pipes and benching in chambers.

The NRA report (1991) revealed that nearly 25% of substantiated incidences of river pollution involved oil. The material is highly visible, and thus all industrial discharges to surface waters where there is the slightest chance of oil contamination must consider one of the removal techniques below.

Although oil storage tanks must be installed in bunds, some bunds are of inadequate total capacity when collected rainwater adds to the volume, and in no circumstances should the bund have drain holes! A small hand pump should be used to empty the bund contents, which may require treatment if contaminated.

Oil/Water Separation Methods

Classically, oil/water mixtures can be separated in tanks or interceptors, the water fraction being abstracted towards the tank base for discharge, and the oil skimmed or pumped off. Providing flow rates allow quiescent conditions, and baffles are placed at inlet and outlet, this simple method works well.

Large separators used by oil refineries normally achieve a 500mg/l standard for discharge to estuaries, although partially soluble components such as phenols and benzene derivatives are discharged and the water often has a characteristic odour. By 1995, the Paris/Oslo Convention will have tightened discharge limits to less than l0mg/l total hydrocarbons and this simple method will not be acceptable. Some form of biological treatment, probably activated sludge with phosphorus and nitrogen nutrient addition, will be required to achieve these limits at large oil refineries and depots.

On the more modest level of the garage forecourt or small factory, a simple structure is adequate to achieve the 500mg/l trade effluent discharge Consent limit; this structure normally has three chambers in series. Construction is often below ground and of brick. A lot of grit and silt also enters the first chamber when draining paved areas, and the position of the interconnecting pipes is as important as regular cleaning out.

Ventilation is essential where petrol spillages may also be contained in the device. Packaged versions are also available in glass reinforced plastic (GRP) from Klargester Environmental Products Ltd., and are claimed to reduce hydrocarbon levels to 5 mg/l through the inclusion of a coalescer unit. This standard is increasingly sought by regulatory authorities where the discharge is to a river or stream.

Tilted plate separation is another useful technique for treating oily water containing 10- 50 mg/l oil, and offers a considerable space saving over large settling tanks. A modular and expandable layout, with closely spaced parallel corrugated plates inclined at 45° or more. Oil globules move upwards and coalesce, form larger droplets and rise to the surface, while sediment slides down the plates. Laminar flow conditions are essential through the plates. More information is available from Environmental Engineering Ltd, and this manufacturer also supplies separators capable of treating the first polluting flush of storm flows from hard standing areas, which are often heavily contaminated with oil products.

Oils, fats and grease from food industry processing are better separated by dissolved air flotation.

CONCLUSIONS

Many of the operating problems to be found on sewage works and industrial effluent plants arise from blockages and loss of treatment capacity caused by detritus, grit and fat deposits. Some form of preliminary treatment is essential. Attention to the capacity and detail of this relatively simple aspect of wastewater treatment will reap dividends in terms of operating reliability, maintenance, effluent quality and the manpower needs of the whole treatment site.

ELEMENTS OF WATER SCIENCE

Water is a vital commodity with many remarkable and unique properties. In his elementary textbook on General Chemistry, Pauling (1953) states: 'Water is one of the most important of all chemical substances. It is a major constituent of living matter and of the environment in which we live. Its physical properties are strikingly different from those of other substances, in ways that determine the nature of the physical and biological world.'

Water constitutes about 70% of the human body and over 80% of most vegetables. It is the primary biological fluid, being the physicochemical reagent that facilitates the metabolic reactions through which food is converted to energy and new cell material. It is the transport medium that removes waste metabolites from the body and it is the medium that cools the body through the thermodynamic mechanisms of transpiration, perspiration and respiration.

Water plays an essential role in photosynthesis; this is the process in which water and carbon dioxide are, by the energy of the sun, converted to carbohydrates in green plants, releasing oxygen at the same time. Photosynthesis maintains the level of oxygen in the atmosphere by replacing that consumed by all living organisms; it is also an important provider of dissolved oxygen to rivers and lakes, where it replaces that used up in the biological degradation of organic matter, thus facilitating the process of self-purification.

Because of its remarkable solvent properties and its ability to entrain and transport particulate matter, water is very susceptible to contamination. It is universally used, domestically and industrially, as a transport medium for waste, generating as a consequence large volumes of waste water.

This textbook outlines the science and technology of the individual processes that are used for the treatment of water to upgrade its quality to meet specific water quality standards. This 'unit operations' approach emphasizes the common process fundamentals whether used in drinking water production or wastewater treatment systems.

Water is a Carrier

The simplest impurities in water are those which are merely carried in it, such as particles of sand. This suspended matter can be filtered out quite easily or (given time) a large part of it may settle out, though very fine or light particles may take an uneconomical time to do so.

Particles too small to be seen under a microscope are called colloids. These are so small that the constant movement of water molecules is enough to keep them floating and distributed through the liquid; they never settle out without assistance. Colloids are roughly in the size range 0.005 to 0.2 microns (a micron is 1/1000 mm). Such very fine particles cannot be filtered by conventional methods; any filter which retains them must be so fine that it raises a very high pressure loss and quickly tends to clog.

The lower size limit of colloids is an arbitrary thing, but we can consider any particles smaller than that to be in solution. In fact there are certain enormous molecules of organic matter occurring in natural waters which are quite as big as colloids. These are sometimes called macromolecules rather than colloids, but this is an exercise in name-calling from which we actually learn nothing.

Water is a Solvent

Once we have passed through this rather uncertain size range, we come to materials which are truly in solution, like salts or detergents. They may be organic or inorganic, and they may be natural or come from some human activity. It is these which give different waters their main properties.

Apart from sea water and brackish water, there are basically three different kinds of waters:

Upland river waters : A glass of water drawn from the tap in Glasgow comes out with a yellowish tinge and a slight 'head' as though it was very thin beer. This is water which has run down the surface of hills over impervious rock, but through and over layers of peat, which is decaying vegetable matter. As a result it has picked up very little inorganic matter and it is soft but, on the other hand, it is full of dissolved and colloidal organic materials which have been picked up from the peat. These cause the yellowish tinge and the surface tension effect which makes the water foam. There are so many organic compounds in any one water that no one has ever analysed and isolated them all. As a group, they have a number of characteristics, some rather harmful, which we will consider later in the book. Because they are not clearly identifiable, the whole business of removing them from water calls for great skill. Wet, hilly, cool places such as Scotland, Wales and Canada tend to have waters like this.

Well waters : Waters which come from deep in the ground (and these include some spring waters) have percolated through mineral layers. These waters contain little organic matter, which the percolation tends to filter out. On the other hand, in passing through the underground minerals, the waters pick up inorganic salts and they tend to contain large quantities of calcium and magnesium, which makes them hard. Whereas upland river waters tend to be corrosive to steel, well waters tend to form scale in pipes and boilers.

Re-used water : A densely populated and industrialized country like the UK tends to re-use its waters. More and more of our supplies come from lowland rivers which contain the effluents from human activities upstream. The commonest of these is domestic sewage.

The inorganic content of treated sewage is 50 to 100 ppm higher than that of the water supply to the township, so sewage not only adds a completely new class of organic matter to a river, it increases the contents of inorganic salts. Much of this additional inorganic matter is due to salts of sodium, which do not increase the hardness but do increase the total dissolved solids. There is also quite a lot of phosphate, which comes from domestic detergents.

The organic pollution from treated sewage will completely decompose, given enough time and dilution in the river. More and better sewage works will do much to reduce the organic content of the water by the time the next user receives it. On the other hand, sewage treatment does nothing to reduce the inorganic content and we can therefore expect the total dissolved solids in our raw waters to go on going up, as more water has to be re-used.

At the same time we already have to use many badly polluted waters; some industries actually use treated sewage directly, and some rivers are not much more attractive as water sources. It is in handling polluted waters that some of the most difficult treatment problems arise.

Water Supports Life

All waters are full of living organisms, spanning the size range from viruses to salmon. Except in the food industry, these are not often of great importance for industrial use but, where harmful, have to be neutralized for domestic consumption.

Water is Cheap

Water's most outstanding property is its cheapness. No discussion of water should ever neglect this.

The most expensive drinking water now being widely distributed in the UK costs industry about 35p per m^3. The worst quality water which is distributed in the UK contains little more than 500 ppm total dissolved solids. We might do better to think of these figures another way: water is a chemical which is being sold biologically safe, better than 99.95 per cent pure, at 35p per tonne (delivered).

Water treatment processes, too, are cheap. The most expensive process in water treatment is for industry and is total demineralization, which yields a product of less than 0.1 ppm total dissolved solids, which is 99.99999 per cent pure. The cost of this remarkable degree of purification, including amortization of plant, might be about another 50p per tonne.

No other process industry uses so cheap a raw material and produces a high purity chemical at so fantastically low a cost. It is this, rather than any other factor, which makes water treatment so specialized a technology.

The Users

There are two broad bands of water users, domestic and industrial, and it will surprise many people to know that the industrial use of water far outstrips the domestic use. The existence of two main types of user leads to problems with definitions. For instance a power-station chemist responsible for supplying pure water to his boilers has a concept of what is pure very different from that of his local environmental health officer who can be called to check pure water supplied for drinking purposes.

With increasingly high standards of purity being imposed upon the municipal authorities for drinking water and increasingly tight consent parameters being imposed upon both municipal sewage and industrial effluent discharges, the cost of using water is increasing dramatically. The economics of water management both municipally and industrially are changing and the manager who ignores his cost/use options may do his organization an expensive disservice.

The types of water treatment available are applicable in most instances to either municipal or industrial supplies but volumetric requirements are widely different and the municipal industry has tended to be dominated by civil engineers who have to store and distribute over wide areas large volumes of water. Industrial users have tended to use chemists for controlling their water requirements because they operate in a smaller geographical area with lower volumes and tend to need a different kind of purity.

The Cycle

All water in a system starts somewhere and ends somewhere. During its voyage it may go through phase changes, become part of another product, be contaminated, cleaned, re-used or discarded. The cycle will normally figure somewhere in the following flow chart.

The two most readily defined waters which are used for potable and industrial purposes are borehole and surface waters. Borehole waters are abstracted from shallow or deep beds and their characteristics derive from their geographical position relative to saline waters, natural aquifers and industrial and agricultural discharges. In some areas there is an unacceptable level of dissolved ferrous and manganese hydroxides which can precipitate on contact with air. In areas of East Anglia and Staffordshire (amongst others) some borehole waters contain unacceptable levels of nitrates. In most areas borehole waters tend to be high in calcium and magnesium hardness. On the whole, however, their treatment for potable purposes requires a lower level of capital expenditure than the surface waters which are subsequently discussed. This economic fact of life has been considerably disturbed by standards imposed within the EC which must lead to considerable expenditure on water supplies to remove impurities which a decade ago were not considered a health risk.

WATER CHEMISTRY

Structure and composition of water

The experiments of Cavendish and Lavoisier in the 1780s established that water is composed of hydrogen and oxygen in the ratio of two volumes of hydrogen to one volume of oxygen, *i.e.* the molecular formula for water is H_2O with a molecular weight of 18. Isotopic analysis (Eisenberg and Kaufman, 1969) has indicated, however, that many natural waters contain a very small rare isotope fraction which generally does not exceed 0.3%. (There are at present three known isotopes of hydrogen, ^{1}H, ^{2}H (deuterium), ^{3}H (tritium), and six of oxygen, ^{14}O, ^{15}O, ^{16}O, ^{17}O, ^{19}O, and ^{19}O,. Tritium is radioactive with a half-life of 12.5 years; the isotopes ^{14}O, ^{15}O, and ^{19}O, are also radioactive, but are short-lived and do not occur significantly in natural water.)

The water molecule may be described as a 'bent' molecule with an atomic configuration, as shown schematically in figure 1.2. In simple terms, its electronic configuration may be described as consisting of four electron orbitals in which four pairs of the outer electrons (six from oxygen and one each from the two hydrogens) are likely to be found. These orbitals are arranged approximately along the directions joining the oxygen atom to the corners of a tetrahedron. Of the four electron orbitals, two are used for the covalent O—H bonds and the remaining two are free orbitals for the remaining so-called lone electron pairs. This electronic configuration confers on the water molecule the important characteristic of charge separation *i.e.* the resultant centres of positive and negative charge do not coincide. The electrically neutral water molecule can thus be considered as an electric dipole, which would be subject to an orientating moment in an electric field. The moment of the dipole is defined as the product of the electric charge by the distance between the charge centres (units esu. cm). In the gas phase the dipole moment of water is 1.87×10^{-18} esu.cm and, owing to molecular association, it becomes somewhat greater in the liquid phase.

Molecular association, or 'hydrogen bonding', which occurs characteristically in ice crystals and is also thought to exist in liquid water, is a further important consequence of the charge separation on the water molecule. Because of this separation the oxygen atom of the water molecule exerts electrostatic attractive forces, in directions corresponding to the orbitals of its lone electron pairs, on the oppositely charged hydrogen atoms of other water molecules. This electrostatic bonding is referred to as hydrogen bonding and, because of it, the structure of ice is arranged in an ordered lattice framework, in which each oxygen atom is tetrahedrally surrounded by four other oxygen atoms. In between any two oxygen atoms is a hydrogen atom which provides the hydrogen bonding. Such a structure of associated water molecules contains interstitial regions which are larger than the dimensions of the water molecule and hence permits the coexistence of free unassociated water molecules.

In liquid water the ordered molecular structure of ice is considered to be partly retained (Pauling, 1960). Its structure may be visualized as consisting of clusters of associated molecules, hydrogen bonded in a tetrahedral framework, plus free unassociated molecules, with continuous interchange between the two. The hydrogen bond is weak relative to the covalent O—H bond. It has a bond energy of 4.5 kcal $mole^{-1}$ as compared with a value of 110 kcal $mole^{-1}$ for the O—H bond. For this reason, the hydrogen bond is easily broken.

The water molecule dissociates to form hydrogen (H^+) and hydroxyl (OH^-) ions. As will be discussed later, these ions play a major role in aqueous chemical reactions. In combination with the polar nature of the water molecule they are responsible for the remarkable solvent properties of water.

Dissolved Substances in Water

Substances dissolved in water (solutes) are dispersed in either molecular or ionic form. The concentration of such substances can be expressed in a variety of units (see Table 9).

Table 9

Basis	*Unit*
Weight/volume (w/v)	mgl^{-1}, kgm^{-3}
Weight/weight (w/w)	$mgkg^{-1}$, ppm
Molarity	moles l^{-1}
Molality	moles kg^{-1}
Normality	equivalents l^{-1}

In many water and wastewater treatment situations, the solution density is effectively the same as pure water (in practical engineering computations taken to be 1000 kg m^{-3}). Hence, the numerical value of the solute concentration, whether expressed as mgl^{-1}, ppm or mg kg^{-1}, is effectively the same. By common usage and method of measurement, the w/v form as mgl^{-1} is preferred for waters and wastewaters, while the w/w form is commonly used in practice in relation to concentrated suspensions such as sludges.

The designation of solute concentration in molar units is particularly appropriate for stoichiometric computations relating to aqueous chemical reactions. A one molar solution of a substance contains one mole of the substance per litre of solution, a mole being the mass of a substance equal to its molecular weight in grams, *e.g.* a mole of oxygen has a mass of 32 g.

The designation of solute concentration in normality terms is convenient for chemical reactions relating to

(1) ion charge,

(2) acid-base reactions that involve hydrogen or hydroxyl ion transfer, and

(3) oxidation-reduction reactions which involve electron transfer. A one normal solution of a substance is a solution that contains one 'equivalent weight' of substance per litre of solution. The equivalent weight of a substance is its molecular weight divided by a factor, the value of which is determined by whichever one of the above three chemical environments exists, as illustrated by the following examples:

(1) **Ion charge.** Iron precipitates from an aqueous solution through the reaction

$$Fe^{3+} + PO_4^{3-} \rightarrow FePO_4 \text{ (s)} \quad (1.1)$$

The ferric iron ion has three charges; hence the equivalent weight of iron is its molecular weight/3

= 55.8/3 = 18.6 g. If the iron concentration in solution is 50 mgl^{-1}, then the normality is calculated as follows:

$$\text{normality} = \frac{50 \times 10^{-3}}{18.6}$$

$$= 2.69 \times 10^{-3} \text{ eq } l^{-1} = 2.69 \text{meq} l^{-1}$$

(2) **Acid-base reactions :** Phosphoric acid dissociates in solution according to the reaction

$$H_3PO_4 \Leftrightarrow 2H^+ + HPO_4^{2-} \qquad (1.2)$$

Two hydrogen ions are released and hence, the equivalent weight of phosphoric acid is its molecular weight/2 = 98/2 = 49 g. If the concentration of phosphoric acid in solution is 50 mg/l, then the normality of the solution is calculated as follows:

$$\text{normality} = \frac{50 \times 10^{-3}}{49} = 1.02 \times 10^{-3} \text{ eq} l^{-1}$$

$$= 1.02 \text{ meq } l^{-1}$$

(3) **Oxidation-reduction reactions :** Ferrous iron is oxidized to ferric iron according to the reaction

$$Fe^{2} \rightarrow Fe^{3+} + e^{-} \qquad (1.3)$$

One electron is released and hence the equivalent weight of iron is equal to its molecular weight. If the concentration of iron is 50 mgl^{-1}, then the normality of the solution is calculated as follows:

$$\text{normality} = \frac{50 \times 10^{-3}}{55.8}$$

$$= 0.9 \times 10^{-3} \text{ eq} l^{-1} = 0.9 \text{ meq} l^{-1}$$

Organic Matter in Water

The concentration of organic matter in water is commonly expressed in terms of the oxygen or equivalent oxygen consumed in its oxidation to carbon dioxide, *i.e.* in terms of oxygen demand.

The biochemical oxygen demand, or BOD_5, is defined as the oxygen uptake in microbial respiration over a 5-day period during which the water sample is incubated out of contact with light, in a constant temperature environment of 25°C. Essentially, BOD is an indirect measure of biodegradable organic matter.

The chemical oxygen demand, or COD, is the oxygen-equivalent of the dichromate consumed in the chemical oxidation of organic matter to carbon dioxide. In the dichromate oxidation-reduction reaction, hexavalent chromium is reduced to trivalent chromium with the loss of three electrons. The equivalent amount of oxygen is computed on the basis that each mole of oxygen accepts four moles of electrons in its reduction to carbon dioxide. The COD is an indirect measure of the concentration of chemically oxidizable organic matter present in water.

Other measures of organic matter in water that are commonly used in water and wastewater treatment processes discussion are total organic carbon, or TOC, and dissolved organic carbon, or DOC. These are derived from direct instrumental measurements of the amount of carbon dioxide produced through the oxidation of the organic carbon present in a water sample.

Water-Dispersed Particles

Water-dispersed species range in size from *true solutes*, *i.e.* ions and molecules to suspended particles visible to the naked eye. Figure 1.3 categorizes the more significant water-dispersed species by size range.

Ions and inorganic molecules may be regarded as true solutes, ranging in size up to about 10 Å or 10^{-3}µm. Low molecular weight organic molecules such as glucose (MW = 180) also belong in this category. High molecular weight organics such as proteins and polymers are an order of magnitude larger and fall within the size range 10^{-3} -10^{-2}µm.

Dispersed particles that are smaller than about 1 um and larger than true solutes are designated as colloids and exhibit characteristic colloidal properties. As indicated in Figure 1.3, they can be observed only by means of an ultra microscope or an electron microscope.

Although, colloids are too small to be distinguished by the naked eye, when viewed at right angles to a light beam, however, colloidal outlines can be distinguished (Tyndall effect). As briefly discussed in section 1.3.5, colloidal particles are subject to random Brownian motion, having a mean velocity that conforms approximately. It follows from this equation that the average kinetic energy of colloidal particles ($\frac{1}{2}mu^2$) is independent of their size.

Thus, the mean velocity of the smallest colloidal particle ($10^{-3}\mu m$) is calculated to be about $100 ms^{-1}$, while the velocity of the largest colloidal particle ($1\mu m$) is calculated to be only about $0.004 ms^{-1}$. Colloidal dispersions have a high degree of stability and are not amenable to removal by sedimentation or conventional filtration processes. The nature of colloidal stability and the processes used to bring. The uniform kinetic energy of colloidal particles, resulting in the very small colloidal particles having very high mean velocities, is a critical factor in the destabilization process.

There is no direct method of measuring the concentration of colloidal particles in water. Turbidity measurement, however, provides a useful index of colloid concentration. Turbidity or cloudiness in water results from the light-scattering effect of very small particles, the outline of which is not visible to the naked eye. It is measured (APHA Standard Methods, 1992) by reference to a standard aqueous turbidity range produced by the dispersion of formazin in water of zero turbidity. Turbidity is normally expressed in nephelometric turbidity units (NTU).

The dispersed species, known as *total suspended solids* (TSS), may be loosely regarded as those solids greater in size than colloids. The TSS concentration is measured by filtration through glass-fibre filter paper having a pore size in the range 1-1.6μm, the retained residue being dried to constant weight at a temperature of 103-105°C.

The 'fixed' suspended solids concentration is the residue remaining after the dried TSS residue is ignited to constant weight at a temperature of (550 ± 50°C. The loss on ignition at this temperature represents the volatile suspended solids.

Living organisms also constitute an important category of dispersed matter in waters and wastewaters. While the larger algae and protozoa are of a size that can be observed by the naked eye, most microorganisms, with the exception of viruses, can be observed using a light microscope and can be separated from water by membrane filtration techniques. Viruses, however, are an order of magnitude smaller than the smallest bacteria. They can be observed only under the magnification of the electron microscope and require filters in the ultrafiltration range for their separation from waters and waste-waters.

Settleable solids constitute a subset of TSS, being that fraction which is separated to form a visibly defined settled solids residue after settling for 1 h. The settleable solids concentration may be expressed in volumetric (ml l^{-1}) or gravimetric (ml l^{-1}) units. The volumetric measurement of settleable solids is made using an Imhoff cone in which a 1 l sample is allowed to settle under quiescent conditions for a period of 1 h; the settleable solids concentration in volumetric terms is represented by the volume occupied by the settled solids at the base of the cone. The gravimetric concentration is determined as the difference between the measured TSS of a test sample and the TSS remaining in the supernatant after a 1 h period of quiescent settling.

2

THE ION EXCHANGE, FLOITATION AND FILTRATION

INTRODUCTION

Ion exchange is the reversible interchange of ions between a solid ion exchange medium and a solution. Cations may be exchanged for Na^+ or H^+ while anions are exchanged for OH^-. Most exchange media in current use are insoluble synthetic polymer resins, although some naturally occurring minerals such as the greensand zeolites (analcite, clinoptilolite, montmorillonite and others) are also used. Synthetic media possess a very large number of charged functional groups, such as $-SO_3^-$ and $-NH_3^+$, to which are loosely attached (by electrostatic forces) small mobile ions of opposite charge.

These latter ions are exchanged for ions of the same sign (so-called counter-ions) in solution. Ions of the same charge as the functional groups are known as co-ions. Resins are designated as cation exchangers or anion exchangers, depending on the counter-ions they exchange. Some resins can exchange both cations and anions and are termed amphoteric ion exchangers. Depending on bead structure and size, resins may be designated as gel resins, macroporous resins, isoporous resins and micro resins. Gel resins and macroporous resin beads are typically 0.3-1.2 mm in diameter and have a three-dimensional matrix of macromolecular hydrocarbon chains, usually consisting of a copolymer of styrene and divinylbenzene. Gel resins have a homogeneous non-porous structure. They imbibe water, which usually results in a swelling of the resin. Ions diffuse through the imbibed water and are exchanged for the mobile ions attached to the polystyrene chains. The matrix of the porous resins is of the same structure as that of the gel resins but the existence of pores allows a more ready access of water into the beads and thus reduces the resistance to diffusion and enables a higher rate of ion exchange. Micro resins are resins in powder form, used in a somewhat similar manner to powdered activated carbon and also requiring separation by a filtration process. The properties of a typical ion-exchange resin are given in Table.

Principle

Ion exchangers are insoluble granular substances which have in their molecular structure acidic or basic radicals that can exchange, without any apparent modification in their physical appearance and without deterioration or solubilization, the positive or negative ions fixed on these radicals for ions of the same sign in solution in the liquid in contact with them. This process, known as ion exchange, enables the ionic composition of the liquid being treated to be modified without changing the total number of ions in the liquid before the exchange.

The first ion exchange substances were natural earths (zeolites); they were followed by synthetic inorganic compounds (aluminosilicates) and organic compounds; the latter materials are used today almost exclusively under the name of resins. This term has been wrongly extended to cover any kind of exchanger.

They are either in the form of granules, as is usually the case, or in the form of beads.

There are two categories: the resins of the gel type and those of the macroporous or loosely cross-linked type. Their basic structure is identical: the macromolecular structure is obtained in both cases by copolymerization of, *e.g.*, styrene and divinylbenzene. The difference between them lies in their porosity. Their high cross-linking degree increases their mechanical strength to both physical (pressure - negative pressure) and chemical (change in the ionic saturation, or exhaustion, stare) stresses.

Gel type resins have a natural porosity that results from the polymerization process and is limited to intermolecular distances. It is a microporous type structure. Macroporous type resins have an additional artificial porosity which is obtained by adding, a substance designed for this purpose. Thus, a network of large canals known as macropores is created in the matrix. These products have a better capacity for adsorption and desorption of organic substances.

Affinity Differences

All ion exchange operations make use of the fact that the resin would rather be associated with some ions than others. As a general rule, a resin has a high affinity for ions with a large number of electrical charges, that is, it prefers Ca^{++} with its two charges to Na^{+} which has only one. Between ions with the same number of charges, the resin prefers the bigger and heavier ion, and so for instance it prefers Ca^{++} (equivalent weight 20) to Mg^{++} (equivalent weight 12) and it prefers Na^{+} (equivalent weight 23) to H^{+} (equivalent weight 1).

There is therefore a pecking order of affinities, in which the commoner cations can be listed like this:

$$Mn^{++++} > Fe^{++} >> Ca^{++} > Mg^{++} >> Na^{+} >> H^{+}$$

The list distinguishes between large affinity differences and small ones. For example, the affinity difference between Fe^{+++} and Ca^{++} is large because Fe^{+++} has one more charge, while the difference between Ca^{++} and Mg^{++} is small because they have the same charge and the resin only chooses between them by virtue of a small difference in their weight.

By contrast, the affinity difference between Na^+ and H^+ is large, because of their very large weight difference.

These affinity differences are greatest in dilute solution and they become relatively small in strong solution, though the pecking order generally remains unchanged.

It was suggested that an ionized solution was like a wife- swapping party to which only couples were admitted, so that the number of men and women was always equal, and in which the couples split utterly, each partner becoming entirely independent.

Seen in this light, an ion exchange resin might be considered as a bench with girls chained on to it; each of the girls must always be accompanied by a man, but they prefer some kinds of men to others.

Equilibrium

When resin and water come into contact with one another, exchange of ions between the resin and the solution is constantly taking place. If the solution contains, say, Ca^{++} ions and the resin is predominantly in the Na^+ form, then the preference which the resin has for Ca^+ means that it will tend to take up Ca^{++} and allow Na^+ out into the solution. However, the Na^+ in the solution also makes efforts to get back on the resin, and a very small proportion of Na^+ ions actually succeeds. Finally an equilibrium is set up, that is, a state of affairs when the number of Ca^{++} ions going on to the resin is just the same as the number of Ca^+ ions being pushed off by Na^+ ions. The net result of this two-way movement is zero, and so the reaction appears to have stopped.

A dilute solution of Ca^{++} ions (such as might be produced by dissolving $CaCl_2$) being brought into contact with a resin which is entirely in the Na^+ form. The resin has a much greater affinity for Ca^{++} and so at first the exchange goes violently towards getting all the Ca^{++} on to the resin. As this exchange proceeds, however, the concentration of Na^+ ions in the solution must increase correspondingly, for the total concentration of ions in the solution remains unchanged. The rate at which Na^+ ions successfully work their way back on to the resin increases with the number of Na^+ ions in the solution. Finally there are only a very few Ca^{++} ions in the solution competing with a large majority of Na^+ ions and at this point equilibrium is reached. If we now drain the water off this sample of resin, and replace it with a strong NaCl solution, the difference in the resin's affinity for the two cations is reduced: it still prefers Ca^{++} to Na^+ but the Na^+ becomes relatively more competitive. This new condition when a new equilibrium is reached. Because of the reduction in the affinity difference, it is possible to get perhaps one third of the total active groups on the resin back into the Na^+ form, while the resin is in contact with a solution which is still two-thirds Na^+ ions.

Column Operation

These two examples represent the exhaustion and regeneration stages of the ion exchange cycle. In practice, however, the water and regenerant are not stirred with the resin but flow through a column of settled resin. The equilibrium which each individual bead of resin reaches is governed by the same rules but column operation introduces some complications of its own. In the first place, natural waters contain both Ca^{++} and Mg^{++}, as well as some Na^+. Also, a finite time is needed to establish equilibrium between resin and liquid and, with water flowing fast down a column, this cannot always be achieved.

Suppose we have a column of resin entirely in the Na^+ form down which we pass a natural well water having, say, the following cations:

Ca^{++} 225ppm as $CaCO_3$

Mg^{++} 50ppm as $CaCO_3$

Na^+ 25ppm as $CaCO_3$

Total 300ppm as $CaCO_3$.

When the first drop of water contacts the first bead at the top of the column, Ca^{++} and Mg^{++} are rapidly taken up by the resin and the first drop goes on down the column as an almost totally Na^+ solution. Passing on through an all Na^+ resin, this drop is in equilibrium with the rest of the column so no effective exchange takes place below the first bead and a softened water emerges from the bottom of the unit.

This first drop of water left behind Ca^{++} and Mg^{++} on the first bead, in the proportion in which they occur in the raw water. The next drop of raw water now comes into contact with the same first bead. The affinity for Ca^{++} is greater than that for Mg^{++}, and so the resin takes up some more Ca^{++} in exchange for which it rejects Mg^{++} into the water. The resin therefore ends up with a higher proportion of Ca^{++} than the raw water, while the second drop which passes on down the unit contains a high proportion of Mg^{++}. This Mg^{++} enriched water then arrives at the second bead, which is still in the all-Na^+ form and therefore takes up all divalent ions keenly while rejecting Na^+ ions.

The third drop of water finds the first resin bead containing a preponderance of Ca^{++} and therefore in equilibrium with it. It finds the second bead enriched with Mg^{++}, which it pushes off down the column.

The Ion Exchange Zone

In this way an ion exchange zone is built up, which has the following characteristics:

- Above the ion exchange zone, the incoming water and the exhausted resin are in equilibrium with one another, so that raw water passes through it unchanged.

- Below the ion exchange zone, the treated water and regenerated resin are in equilibrium with one another so that the water passes through it unchanged also. All exchange takes place within the zone.

- As water flows down the column and exchange proceeds, the zone moves slowly down the column.

- Inside the ion exchange zone, there is a tendency for less tightly held ions like Mg^{++} to be pushed in front, creating a front of water enriched in Mg^{++} which runs ahead of the raw water.

- This front of Mg^{++} rich water creates a band of Mg^{++} rich resin which moves down the column at the head of the exchange zone.

- At the beginning of the run, the exchange zone is very shallow, but it deepens as it moves down the column.

- When the head of the ion exchange zone reaches the bottom of the unit, the resin bed is normally considered to be exhausted and the unit is taken out of service for regeneration. At this point the breakthrough of hardness is mostly of Mg^{++} ions, because it is the enriched front which is breaking through.

- At this point, only the resin above the exchange zone is totally exhausted: within the exchange zone there is an Mg^{++} rich zone which could remove Ca^{++} in exchange for Mg^{++}, and at the very bottom there is also a small amount of Na^{+} which has not yet had time to come to equilibrium with the advancing front.

This property of taking up ions in bands according to their affinity is called chromatography and is used in some chemical processes for the separation of, for example, rare metals.

Regeneration

When the unit is exhausted, it has to be regenerated with strong NaCl brine and the normal practice is to pass this down the column in the same direction as the water flowed during the exhaustion stage. The brine first contacts a part of the bed which is substantially in the Ca^{++} form, and therefore removes Ca^{++} ions which then pass on down the column. When this Ca^{++} rich solution gets to the bottom of the column, there is some tendency for the Ca^{++} to displace Mg^{++} and even more to displace any Na^{+} which may have remained behind at the very bottom. A regeneration zone is formed which now travels down the column in exactly the same way as the ion exchange zone. It carries in front of it a Ca^{++} rich solution which tends to expel all the less tightly held ions from the resin, though it must be remembered that the affinity differences have been greatly reduced because of the much higher concentration at which regeneration takes place. However, the fact is that pure NaCl brine only comes into contact with resin after a Ca^{++} rich solution has already gone over it, and has left some Ca^{++} on the resin. One important difference between the exhaustion and regeneration stages is that in exhaustion the equilibrium is reached very quickly, while the regeneration equilibrium takes some time. In regeneration, therefore the zones are much less sharply defined.

THE ION EXCHANGE REACTION

Use of a Reversible Reaction of the Softening Type

$$R\text{-}Na_2 + Ca^{2+} \rightleftarrows R\text{-}Ca + 2Na^{+}$$

As is the case with any chemical equilibrium, it is governed by the law of mass action, the reverse reaction of which corresponds to the regeneration of the exchanger.

If the liquid to be treated is brought into static contact with the exchange material, the reaction stops when equilibrium is reached between the liquid and the resin.

Therefore, in order to achieve substantially complete exchange, it is necessary to create successive equilibrium stages by percolating the water through superimposed layers of exchange material.

There is always a varying degree of leakage of the ion that one wants to remove.

Laws governing a reversible ion exchange: for each reaction involving two ions A and B, the equilibrium between the respective concentrations A and B in the liquid and in the ion exchange substance can be shown graphically.

Under conditions of equilibrium, and for a concentration B of X% in the solution, the exchange material becomes saturated up to a concentration of Y%, When the two ions A and B have the same affinity for the exchange material, the equilibrium curve corresponds to the diagonal of the square. The more marked the exchange material's preference for ion B, the further the curve moves in the direction of the arrows.

The form of the curve for a given system of two ions depends on a number of factors: nature and valency of the ions, concentration of ions in the liquid, and type of exchange material. In a system $Ca^{2+} \rightleftarrows 2Na^{+}$ applied to a sulphonated polystyrene, the exchange material always has a greater affinity for calcium than for sodium, and the more dilute the solution the more this will be the case. As mentioned above, static batch treatment effected by bringing the liquid and the exchange material into contact in a tank would reach a certain point on the curve and remain there.

If the treatment is to be continued until one ion is effectively removed in favour of another, the point of equilibrium must be progressively shifted by passing the liquid through a series of successive layers of the exchange material containing fewer and fewer ions to be retained, thereby moving along the equilibrium curve almost to the zero concentration point for the unwanted ion. If we take a layer of exchange material entirely in form A, and if a liquid containing ion B is passed through it, the successive equilibrium points between A and B give a series of isochronous concentration curves for two ions of similar affinity and by where the èxchange material has a much greater affinity for ion B' than for ion A'. The "leakage point" is reached when the isochronous curve leaves the vertical right-hand axis (positive concentration of B in the liquid outlet curve). The curves for the two cases of different affinity are represented.

If the area B/A+B represents the fraction of the total exchanger capacity used when the leakage appears, it is dear that this fraction is much greater for B' than for B.

This is also obvious with the treated liquid in the form of "the exhaustion curves" The form of the exhaustion curves depends not only on the static equilibrium curve mentioned above, but also on the "èxchange kinetics" between the liquid and the exchange material; these kinetics involve the penetration of solutes into the exchanger, and are governed by laws known as the "Donnan equilibrium laws". These phenomena are very complex; they involve both the degree of dissociation and concentration of ions, the temperature, the nature of the exchange material/liquid interface, and the kinetics of penetration into the solid that constitutes this exchanger.

The total capacity of an exchanger, *i.e.*, the total number of equivalents available for exchange per litre of exchange material, is only of very relative practical value; for commercial application it is the "useful capacity" defined from the isochronous graphs or exhaustion curves described earlier which is of importance. Accepting the existence of an exchange material-liquid equilibrium curve, the quality of the liquid treated by an exchange layer depends on the quality of the last layer through which the liquid passes, whatever the qualities of the preceding layers.

If we consider a reversible reaction of the type:

$$R - A + B \rightleftarrows R - B + A$$

that represents "service" from left to right and "regeneration" from right to left, it is necessary to examine the state of the exchanger at the start of a treatment cycle following a regeneration cycle. It is clear chat at the beginning of the treatment cycle the quality of the treated water, characterized by the ion leakage, will essentially depend on the degree of regeneration of the last layer of exchange material.

These factors must be borne in mind during the examination of the different qualities of exchange materials and their industrial applications.

Non-reversible Reaction

This applies to the removal of a strong acid by a strong base anion exchanger:

$$HCl + R - OH \rightarrow R - Cl + H_2O$$

The reverse reaction (hydrolysis) is virtually non-existent; the exchange is complete and can be obtained just as well under static contact or percolation conditions. In this case ion leakage may be zero, provided that the contact time between the water and the resin is long enough.

Equilibrium reactions that give rise to an insoluble compound can be likened to this type of exchange. For example, if sea water is treated with an exchanger saturated with silver ions, the following reaction is obtained:

$$R - Ag + NaCl \rightarrow R - Na + AgCl$$

Since AgCl is insoluble it precipitates. Under these conditions, and according to Berthollet's law, the equilibrium shifts completely and the reaction is complete, even under static conditions.

Ion exchange is not instantaneous, and the rare of reaction depends upon the type of resin. In practice this type of exchange gives exhaustion curves similar to those in Figure.

The two types of reaction mentioned above can be used:

- to remove one or more unwanted ions from the liquid under treatment,
- to select and concentrate in the exchanger one or more ions that will be bound later in the purified and concentrated state in the regeneration or elution liquid.

Use of a Previously-attached Complex Anion

This complex ion is liable to cause secondary reactions, for instance oxidation-reduction phenomena affecting the ions in the water or liquid to be treated, without itself dissolving in the liquid.

For example: absorption of dissolved oxygen by oxidation of a sulphite anion resin $R - NH_3 - HSO_3$ into a sulphate resin $R - NH_3 - HSO_4$.

Other Uses

The ion exchange laws do not concern the use of ion exchangers for other purposes (catalysis, adsorption).

Methods of Regeneration

In the case of softening and demineralization processes, the end of the cycle is reached when the exhaustion curve corresponds.(compounds A' and B'). It can then be assumed, at least as far as the upper layers are concerned, that the ion exchanger is saturated with B' ions and is in equilibrium with B' concentration in the inflowing solution.

Regeneration is carried out by causing a concentrated solution of A' ions ro flow through the exchanger either in the same direction as the exhaustion (cocurrent regeneration) or in the opposite direction (countercurrent regeneration).

- **Cocurrent regeneration** : In this operation the concentrated solution of A' ions is initially brought into contact with those layers of the ion exchanger saturated with the B' ions which are expelled from the resin; these B' ions are then carried to those layers of the ion exchanger which are at a lower level of exhaustion and where the conditions are favourable ro their capture; during the first stage of regeneration it is therefore mainly the A' ions which are eluted from the column.

  Therefore it appears that, in order to achieve total regeneration of the ion exchanger, it is necessary to subject a quantity of ions corresponding to the ratio A'/B' to a double exchange process.

  Finally, if the quantity of regenerating solution is limited, the B' ions will not be completely eluted from the ion exchanger and the bottom layers will not be fully regenerated.

  Consequently, during the following cycle, the B' ions will undergo self- regeneration by the A' ions displaced from the upper layers.

- **Countercurrent regeneration** : the course of events will be different when the regenerating reagents are made to flow upwards from the bottom. In this case, the concentrated A' ions first of all encounter the resin layers with a low concentration of B' ions, the elution of which therefore takes place in favourable conditions; what is more, the B' ions cannot be recaptured in the exhausted upper layers

Regeneration therefore takes place in far more reversible conditions than when the cocurrent technique is used, and from a thermodynamic point of view, this means greater efficiency.

Two important advantages of the principle of counter-current regeneration should be emphasized:

- higher efficiency, and consequently reduced reagent requirements, given equal quality,
- better quality of the treated water thanks to the fact char the bottom layers are regenerated with a large excess of reagent.

Ion Exchange Vocabulary

- **Exchange capacity of an exchanger** : this is the weight of ions that can be retained per unit volume (or sometimes per unit weight) of the exchange material concerned. The capacity is expressed in gramme-equivalents or in degrees per unit volume of compacted resin. A distinction is drawn between :
- coral capacity, which is the maximum volume of ion that can be exchanged, and which characterizes a given resin, regeneration.
- breakthrough capacity, which is the usable fraction of the above, depending on the hydraulic and chemical conditions of each individual application.
- **Bed volume** : $\frac{\text{volume per hour of liquid to be treated}}{\text{volume of resin}}$
- **Ion flux** : bed volume x salinity of water
- **Regeneration level** :

$$\frac{\text{weight of reagent used}}{\text{volume of the ion exchange material}}$$

- **Regeneration rate** :

$$\frac{\text{gramme} - \text{equivalents of regenerating reagent}}{\text{gramme} - \text{equivalents of eluted ions}}$$

This ratio is always equal to or greater than 100% (100% corresponds to the Stoichiometric efficiency).

- **Regeneration efficiency** : this is the opposite ratio to the above.

- **Ion leakage:** this is the concentration of the unwanted ion left in the treated liquid. It is expressed in mg.l^{-1}, μg.l^{-1}, milliequivalents/litre, sometimes in % in relation to the inflowing liquid.

- **Breakthrough** : this is the maximum permissible ion leakage requiring the production cycle to be shut down.

- **Attrition** : mechanical wear of the exchanger grains as they are being used.

MAIN TYPES OF ION EXCHANGERS

Properties of an Ion Exchange Material

An ion exchange material for industrial use must meet the following requirements :

- *The product must be insoluble under normal conditions of use.*

  In practice, all the exchange materials in current use meet this requirement, and their true solubility at ambient temperatures is nor detectable by the usual methods of analysis under normal conditions of flow and temperature. This is no longer true of certain exchange materials once a certain temperature is reached.

- *The product must be in the form of grains of maximum homogeneity and of dimensions such that their head loss in percolation remains acceptable.*

  Ion exchange materials used for the applications described below take the form of grains 0.3 to 1.2mm in size. Powdered resins of between 5 and 30 microns, known as "microresins", are available for certain special uses (treatment of condensates, waters from nuclear systems).

- *The changes in the state of the exchange material must not cause any deterioration in its physical structure.*

  The exchange material may be required to retain ions or ionized complexes of highly varied dimensions and weights.

  In some cases this causes an obvious contraction or swelling (up to 100% for some carboxylic resins (HCO_2-R) between the H and NH_4 phases). This swelling and contraction, obviously, should not cause the grains to burst. In the most difficult cases the design of the apparatus should allow for this expansion without causing excessive stresses in the bed.

It should never be forgotten that there are certain (all too often disregarded) limitations on the use of ion exchangers :

- ion exchangers can function only in the presence of a liquid phase of limited concentration,
- ion exchangers are made to retain ions and not to filter suspended solids, colloids or oily emulsions. The latter substances can only shorten the life of the exchange materials,
- the complex problem of removing soluble organic substances must be the subject of a detailed study,
- the presence of large quantities of dissolved gases in the water can cause serious disturbances in the activity of the exchangers,
- the powerful oxidizing agents Cl_2 and O_3 affect some resins,

- generally speaking, great caution should be used in the practical application on an industrial scale of laboratory results, and also when reading the documentation produced by ion exchanger manufacturers.

CONVENTIONAL TECHNIQUES

General

It is important to emphasize that the techniques related to ion exchange processes should not be used unless the raw water has been subjected to a form of preliminary treatment suited to its type, which must include the removal of suspended solids, organic matter, residual chlorine and chloramines, etc. The preliminary treatment varies with the type of ion exchanger used.

The conventional systems are made up of fixed-bed ion exchangers-regenerated on a cocurrent basis: the water to be treated as well as the regeneration solution pass through the resin bed from top to bottom.

The complete cycle of exchange consists of the following phases:

- **service** (or production): the operating cycle of an exchanger is determined by the exchange capacity of the layer. This corresponds to an exchangeable mass of ions, and consequently, to a certain volume of water treated between two regeneration operations,

- **loosening** : an upward flow of water loosens the resin bed and provides for the removal of particles and resin debris that may have risen up to the surface,

- **regeneration** : the diluted regenerant passes slowly through the resin bed from top to bottom,

- **displacement** (or slow rinse): water is introduced at the same rate of flow and in the same flow direction as the regenerant until nearly all of the latter is washed our and removed,

- **fast rinse :** water is introduced at the production rate of flow until a quality of treated water is obtained that is suitable for use in the production process.

- **Description of an ion exchange unit :** whatever the type of exchange, whether for softening, carbonate removal or demineralization, each appliance normally consists of a vertical, closed, cylindrical vessel holding the resin. The latter can be placed in direct contact with the treated water collection system, which may consist either of nozzles evenly distributed over a tray, or of a system of perforated tubes of a suitable number and size. The resins may also be supported by a layer of inert granular material such as silex, anthracite or plastic beads. The layer itself is drained by the collection system.

  Sufficient free space is left above the resin bed to allow it to expand normally (between 30 and 100% of the compacted volume depending on the type of resin) during countercurrent expansion.

  Both the water to be treated and the regenerant are admitted at the top of the vessel by a distribution system of varying complexity.

  The appliance has an external set of valves and pipes for the various operations of service, expansion, regeneration and rinsing. The valves may be manually or automatically controlled, or can even be replaced by a central multiport valve.

Softening

A cation exchanger regenerated with a sodium chloride solution is used for this purpose. All the salts in the water under treatment are transformed into sodium salts. The hardness of the treated water is virtually nil. Its pH and alkalinity values remain unchanged. Softening can be done after preliminary purification by lime, which removes the bicarbonates and reduces the M alk. to a value generally between 2 and 4 French degrees. In this case, the water obtained is both free from carbonates and softened.

Carbonate Removal

This process uses a carboxylic resin which is in the R-H form, having been previously regenerated by an acid. This resin has the property of retaining metallic cations and releasing the corresponding anions in the form of free acid, until the pH of the treated water reaches a level of between 4 and 5, at which point all the carbonic acid from the bicarbonates is released. The cations associated with the anions of strong acids (chlorides, nitrates, sulphates) are not retained by the resin.

Under these conditions, the treated water contains all the original salts of strong acids and an amount of dissolved CO_2 which is equivalent to the bicarbonates in the raw water. The alkalinity of this water may be nil, and its hardness equal to the TH - M alk. value of the raw water; the hardness value may therefore fall to zero if the TH is equal to or less than the M alk., since the alkaline-earth ions are exchanged rather than the alkaline ions. In the opposite case, a zero hardness may be obtained by combining in the same vessel a layer of carboxylic resin and a layer of sulphonic resin, regenerated in turn with a strong acid and a solution of sodium chloride. The carboxylic resin works in the H cycle and retains the TH in a quantity which is equivalent to the M alk. The sulphonic resin exchanges the sodium ions for permanent hardness which is equal to TH-M alk. In this way a carbonate-free and softened water is obtained.

With **water containing sodium bicarbonate**, the efficiency of carboxylic resins is poor, and the **H-Na method** is sometimes used instead: a sulphonic resin in H form is placed in parallel with another in Na form and while the former retains all the cations and releases the corresponding acids, the latter produces softened water. A mixture of decationized and softened water in suitable proportions provides treated water of the same composition as the first method.

However, this method has the disadvantage of requiring the acid water ro be kept strictly proportional to the water containing bicarbonate alkalinity, as otherwise an acid, and thereby corrosive, mixture is obtained.

With these systems it is generally advisable to remove the dissolved CO_2 produced by the ion exchange process.

Demineralization

A number of various demineralization processes have been studied which are essentially based on the following factors: the quality of water to be obtained, the composition of the water to be treated and the consumption of regeneration reagents. Other considerations such as capital outlay, installation, etc., may also influence the composition of the system.

The description given below of the various combinations of exchangers most frequently employed, uses the terminology:

- **WCR** : weak acid cation resin
- **SCR** : strong acid cation resin
- **WBR** : intermediate or weak base anion resin
- **SBR** : strong base anion resin
- **$|CO_2|$** : CO_2 removal
- **MB** : mixed bed.

Partial Demineralization

This is comprised of a unit filled with strong acid cation exchanger (SCR), regenerated by a strong acid, which operates in series with a unit filled with weak (or intermediate) base anion exchanger (WBR), regenerated by caustic soda or ammonia. The water produced is used as is as long as the carbonic acidity is not damaging; otherwise it is deaerated on a CO_2 removal unit (decarbonator) that is placed either upstream or downstream from the anion exchanger.

The treated water contains the totality of the silica present, and when it is deaerated, the level of carbonic acid is approximately 15 mg.l^{-1}; depending upon the regeneration rare adopted for the canon exchanger, the conductivity can range between 2 and 20 μS.cm^{-1}. The pH level is in the order of 6 to 6.5 as long as the removal of carbon dioxide is done correctly. This type of system produces boiler feedwater for medium pressure boilers and water for various industrial processes.

Total Demineralization

- **SCR + SBR systems** : all ions, including silica, are removed. In the majority of cases it is advisable to reduce the flux of ions passed to the anion exchanger by installing, between the anion exchanger and the cation exchanger, a CO_2 removal unit intended to reduce the CO_2 content to a few mg.l^{-1}. This brings about a reduction in the volume of strong base anion resin and in the regeneration reagent requirements. The quality of demineralized water essentially depends on the regeneration race of the cation exchanger. The ion leakage takes the form of a trace of caustic soda (or of lime, if the raw water contains no sodium) from the cation exchanger. A reduction in the level of silica itself depends on the level of caustic soda that remains in the demineralized water. In practice, in most of the cases the water obtained by this method has a conductivity of 3 to 20 mS.cm^{-1}, a silica level of 0.05 to 0.5 mg.l^{-1} and a pH ranging between 7 and 9.

This is the simplest arrangement and a demineralized water that may be used in a wide variety of applications can be obtained with it.

- **SCR + WBR + SBR system :** this combination is a variation of the previous one. It provides exactly the same quality of water, while offering economic advantages in the case where the water to be created contains a high proportion of strong anions (chlorides and sulphates).

  In this system, the water, after first passing through the weak base anion exchanger, in turn passes through the strong base anion exchanger. The optional CO_2 removal unit may be installed either between the cation exchanger and the first anion exchanger, or between the two anion exchangers.

  The regeneration of the anion exchangers takes place in series with the caustic soda solution first passing through the strong base resin and then through the weak base resin. This method requires the use of much less caustic soda than was used in the previous one, because the excess caustic soda remaining after normal regeneration of the strong base resin is usually sufficient to regenerate the weak base resin completely.

  Moreover, when raw water contains a high proportion of organic matter, the weak base resin protects the strong base resin.

- **Systems with WCR + SCR grouping :** this combination is advantageous in cases where the water contains a high proportion of bicarbonates. In this system, regeneration is effected in series first passing through the sulphonic exchanger and then through the carboxylic exchanger.

  Since the carboxylic resin is regenerated more or less stoichiometrically, from the excess free acid that remains after the regeneration of the sulphonic resin, the total regeneration rate is considerably lowered

- **Mixed bed installations (MB)** : this process differs essentially from the separate bed system in that the two strong resins, the cation and the anion, are joined in a single vessel. The two resins are intimately mixed by agitation with compressed air. The grains of resin are thus arranged side by side, and the whole bed behaves like an infinite number of anion and cation exchangers in series.

To carry out regeneration, the two resins are separated hydraulically during the loosening phase. As the anion resin is the lighter, it rises to the top, while the heavier cation resin falls to the bottom.

When the resins have been separated, each of them is separately regenerated in turn with caustic soda and a strong acid. Any excess regenerant is removed by rinsing each bed separately. After partial emptying of the vessel, the two resins are remixed with compressed air. Rinsing is completed and the vessel is then ready for a fresh cycle.

The advantages of mixed bed systems as compared with separate bed systems are as follows:

- the water obtained is of very high purity and its quality remains constant throughout the cycle (its conductivity is below 0.2 $\mu S.cm^{-1}$, its silica level is less than 20 $\mu g.1^{-1}$),
- the pH is almost neutral,
- rinse water requirements are very low.

The disadvantages of mixed bed systems are a lower exchange capacity and a more complicated operating procedure because of the requirement that the separation and remixing processes be carried out absolutely correctly.

Mixed bed exchangers can be used directly on raw water as long as it contains very few ions (water which has undergone prior treatment by reverse osmosis or distillation, condensed water, nuclear pool water in dosed loops, etc.). A complex system of ion exchangers can be replaced by a single mixed bed.

Special layouts have also been used as follows:

- SCR + $|CO_2|$ + MB,
- softener + MB,
- SCR + WBR + $|CO_2|$ + MB: useful arrangement for a water containing many strong anions.

However, the mixed bed exchangers are most often used in polishing treatment.

- **Installation equipped with a polishing system** : the quality of water flowing out of a primary system, whatever its composition, is determined by the ion leakage from the cation exchanger. This ion leakage varies, depending upon the properties of the raw water and the rare of regeneration. The quality of the demineralized water obtained is not sufficient for certain uses such as that of feedwater for very high pressure boilers and various applications in the chemical, nuclear or electronics industries. Therefore, it has to be further treated in a system known as a polishing plant.

The ion leakage from the cation exchanger is converted to a free base on the anion exchanger, which therefore entails a silica leakage from the latter. As a result, a polishing system must necessarily contain a strong acid cation exchanger and a strong base anion exchanger.

The polishing system may be arranged so that there are two columns in series with regeneration taking place in the direction SBR2 - SBR1 and SCR2 - SCR1. In these arrangements the polishing exchangers are perfectly regenerated and the quality of the demineralized water is excellent (the conductivity is less than 1 $\mu S.cm^{-1}$, its silica level ranges between 5 and 20 $\mu.g.1^{-1}$).

This system, however, is being used less and less frequently. Mixed bed exchangers that produce water with a conductivity in the order of 0.05 $mS.cm^{-1}$ at 25°C, and have silica leakages that are considerably below 10 $\mu g.l^{-1}$, are preferred. A polishing mixed bed, due to the low volume of inflowing ions, is regenerated only every 5 to 10 cycles of the primary system. In some applications, it may be sufficient to have as a polisher a weak or a strong acid canon exchanger designed to neutralize the caustic soda leakage from the anion exchanger in the primary system. By using this polishing exchanger, known as a "buffer filter", it is possible to obtain water that is virtually free of cations (with a conductivity below 1 $\mu S.cm^{-1}$) and has a pH value of between 6 and 7.

SIZING A DEMINERALIZATION SYSTEM

The following data are necessary for sizing:

- M alk. of the raw water in French degrees,
- SSA of the raw water in French degrees (SO_4 + Cl + NO_3),
- silica content as $TSiO_2$ (1° = 12 $mg.1^{-1}$ SiO_2).
- content of carbon dioxide, TCO_2, in the water after passing through the cation exchanger and, where appropriate, after CO_2 removal,
- volume V of water to be supplied between regeneration processes, in m^3, including service water if appropriate,

- hourly flow rate Q in m^3,
- exchange capacity C of the resins expressed in degrees-litres per litre of compacted resin (the degrees may be replaced by milliequivalents, where 1 milliequivalent = 5 French degrees).

In the case of conventional systems, the exchange capacity is calculated for each resin, with the aid of information furnished by the manufacturer.

The anion exchanger is calculated first for capacity C : the volume to be used is given by the formulae:

$$V_a = \frac{V \times SSA}{C}$$

in the case of a weak base exchanger, and:

$$V_a = \frac{V \times (SSA + TCO_2 + TSiO_2)}{C}$$

in the case of a strong base exchanger.

The cation exchanger is calculated next, allowing for the additional water αVa necessary to rinse the anion exchanger, where a may vary from 5 to 20 depending on the type of resin. This results in:

$$V_c = \frac{(V + \alpha V_a)(SSA + M\ alk.)}{C}$$

The volumes calculated must then be compared with the hourly flow race to be treated. There are upper limits to the flow rate or to the bed volume.

If V_c or V_a are too low, they should be readjusted, possibly by increasing the cycle volume V.

ION EXCHANGE APPLICATIONS

In the water treatment field ion exchange is used for water softening and demineralization. A common system in use is the down-flow fixed bed arrangement.

In softening plants a strong acid sodium cation exchange medium is normally used, while in demineralizing plants, beds of H^+ and OH^- media may be used in series or in a single mixed bed. To avoid short- circuiting, beds are commonly not less than 0.8 m in depth. The required bed volume for a particular application can be most reliably obtained from the results of a laboratory column study. Flow rates obtaining in practice generally fall within the range 0.2-1.0 $m^3\ min^{-1}\ m^{-3}$ of medium. When breakthrough occurs, the bed is generally back-washed with clean water to remove any accumulated solids before regeneration is carried out.

The influent to an ion exchanger must be low in turbidity and in organics which might be adsorbed by the exchange medium, thus inactivating its exchange capability.

There are three general types of filtration processes, depending on the method used:

- filtration on a support,
- filtration through a granular filter bed,
- filtration with a filter cake.

Filtration Mechanisms

According to the characteristics of the particles to be filtered out and the filtration material used, one or more of the following principal mechanisms can be involved: retention, attachment and detaching.

Retention Mechanisms

There are essentially two types:

- **Mechanical straining:** this retains all particles larger than the mesh size of the filter or the mass of the particles already deposited which themselves form filter material. The finer the mesh of the filtering material, the more marked this phenomenon will be: it is of little significance in a filter bed composed of relatively coarse material, but is of great importance in filtration through a fine-mesh media: strainer, filter sleeve, etc.

- **Deposit on the filter material:** the suspended particle follows a line of current in the liquid; depending on its size in relation to the pores, it may be able to pass through the filter material without being retained. However, various phenomena cause its travel to change and bring it into contact with the material.

The following phenomena can be identified:

- direct interception by rubbing,
- diffusion by Brownian movement,
- inertia of the particle,
- **settling:** particles may settle on the filter material by gravity, whatever the direction of filtration.

These retention mechanisms occur mainly during the process of in-depth filtration.

Attachment Mechanisms

The attachment of particles to the surface of the filter material is promoted by a slow rate of flow, and is caused by physical forces (jamming, cohesion), and by adsorption forces, mainly Van der Waals forces.

Detaching Mechanisms

As a result of the mechanisms referred to above, the space between the walls of the material covered with particles that have already deposited, decreases. Consequently, the velocity of flow increases. The retained deposits may become partially detached and be driven deeper into the filter material or may even be carried off in the filtrate.

The solid particles in a liquid and the colloidal particles that are flocculated to a greater or lesser degree do not have the same characteristics and do not react to the same extent to the above mechanisms. Direct filtration of a liquid in which the suspended solids retain their original state and electrical charge will therefore be very different from filtration of a coagulated liquid.

Clogging and Washing of the Filter Material

Clogging is the gradual blocking of the interstices of the filter material. Clogging causes the head loss to rise.

If a constant intake pressure is maintained, the flow of filtrate will decline. To keep output constant, the initial pressure must be increased as the filter becomes clogged.

The clogging rate depends on:

- the matter to be retained: the more suspended solids there are in the liquid, the greater the cohesion of these solids, and the more liable they are to proliferate (algae, bacteria),
- the filtration rate,
- the characteristics of the filter material: size of the pores, uniform particle size, roughness, shape of the material.

The filter becomes clogged when it reaches the maximum design head loss. It must be restored to its original condition by efficient and economic washing; the method used depends on the type of filter and the type of matter it retains.

Choice of Filtration Method

Several different criteria govern the choice between the different types of filtration on support and filtration through a filter bed:

- characteristics of the liquid to be filtered, its impurities and their evolution with time,

- the desired quality of filtrate and the permissible tolerances,
- the quality of the mass of retained material when the object is to recover it,
- installation conditions,
- facilities available for washing.

In selecting a filter, the possibility of easy, efficient and economical washing is as important as obtaining the best filtration quality, since this quality will only be maintained if the washing process allows the filter material to remain intact.

FILTRATION ON SUPPORT

We can distinguish:

- straining, which is surface filtration in which the mesh openings are relatively large (larger than approximately 30 microns),
- fine filtration, in which filtration through thin support of the coarsest particles occurs together with in-depth filtration of the finest ones.

One aspect of surface filtration should be emphasized: it removes particles that are larger than the filter pores **insofar as they cannot change in shape**. Under pressure and with substantial head losses, it is possible that larger particles can pass through.

Straining and Microstraining

This involves a relatively coarse filtration or filtration through a thin support made of metal or plastic fabric, or filtering elements with regularly shaped pores. Depending on the size of the openings, the process is referred to either as microstraining or as macrostraining, of which the latter is covered in the area of fine screening.

Mesh or pore openings	**25-150 μm**	**0.2-4 mm**	**2-6 mm**
Method	Microstraining	Macrostraining	Fine screening
Operation	Gravity or under pressure	Gravity	Gravity

The "removal capacity" is denned by the mesh void: the system retains all particles which are bigger than the mesh size. During the operation, strained particles may partially obstruct the mesh, and the filter may retain particles that are smaller than the size of the actual removal capacity.

In gravity operation, the maximum head loss designed for these filters is usually low—in the area of several dozen centimetres. This is due to the fragility of the cloth used, which may tear under the pressure applied and/or when washed.

Free Surface Microstraining

The main objective of microstraining is to remove the plankton from surface waters. This process will, of course, also remove suspended solids of large size and plant or animal debris from the water. It may also be used after biological purification or lagooning to remove residual suspended solids.

Optimum efficiency is obtained by maintaining a more or less constant total head loss that results from partial clogging by the particles to be retained. However, the efficiency of such an installation will always be limited by a number of factors:

- the washed filter cloth does not carry an effective deposit at the start of the filtering cycle, and filtration is then limited to the size of the mesh alone,
- plankton removal is never complete. The plankton can grow again, particularly when the temperature rises,

- certain very small eggs can easily pass through the filter cloth and hatch in the downstream tanks, where crustaceans visible to the naked eye may develop,
- because of the risk of corrosion of the microstrainer cloth or its supports, it cannot be used for continuous treatment of heavily prechlorinated water,
- microstrainers have to be fairly large to cope with peaks of plankton growth which occur several times a year. If they are too small, the output of the plant could be reduced significantly during these peaks and during alluvial high-water periods.

The smaller the mesh, the more important the straining area has to be. Thus, with a 35-micron mesh size, the filtration rate should be 35 $m.h^{-1}$ at the most, calculated over the total area of the strainer (50 $m.h^{-1}$ on the real submerged area).

The reduction in clogging capacity of the water by microstraining ranges from 50 to 80% with an average of about 65%. As a comparison, a good settling rank gives a reduction of 80 to 90% without prechlorination and 95 to 99% with pre- chlorination. The microstrainer should be used only for water containing few suspended solids. It has no effect on colour and on dissolved organic matter, and only removes the coarsest proportion of the suspended particles.

For a really effective disposal of plankton, clarification preceded by oxidation is essential.

Microstraining under Pressure

In industry, water may be strained under pressure. The aim is to assure:

- either protection against the clogging of relatively large openings (several mm) with a straining threshold of 0.15 to 2 mm. This is the case with once-through or open recirculating cooling systems,

- or continuous removal of fine solids. The straining threshold can be lowered to 50-75 microns, if not lower, and the unit becomes part of a treatment line. This may be the case where sea water is injected to obtain secondary recovery of oil.

These filters, used mainly for filtering industrial water, are called automatic regeneration rotary filters or, sometimes, pressure mechanical filters. They are used with differential pressures of 0.5 to 2 bar. The initial head loss should be small (0.15 to 0.5 bar).

Filtration using Cartridges and Candles

The Goal

In the treatment of water, filters are used to solve one of the following problems:

- **Very high quality of the filtrate** from water containing a very small amount of particles and without the release of support:
  - condensates from HP boilers, whether starting up or in steady operation,
  - the feeding of systems with ultrapure water,
  - protection of reverse osmosis membranes.

  The aim is to remove increasingly finer particles down to 0.5 microns (bacteria). Disposable supports may be used despite their cost.

- **Protection of hydraulic systems** preventing the carrying away of discrete particles:
  - injection of sea water for secondary recovery,
  - cooling systems or process water systems.

It is necessary to prevent the loss of all filtering materials (fibres, resins, activated carbon, etc.) and any input of particles by the air. The level of filtration ranges from 10 to 200 microns. The use of regenerable supports is now in the preliminary phase.

Choice of Media

This depends on the desired efficiency and the parameters of use.

- **Criteria for filter efficiency**
  - **The nominal removal threshold** has been determined by the manufacturer for a given support and suspension to be treated. It corresponds to the dimensions of the finest particles retained without giving a strict removal percentage.
  - **The "absolute" removal threshold** is the diameter of the smallest particle for which the β index reaches the value desired for a given application. This β index, *i.e.*:

$$\frac{\text{Number of panicles in the water to be filtered}}{\text{Number of particles in the filtrate}}$$

is measured by the filtration of a reference suspension made up of silica powder ACFTO for suspended solids larger than 1μm, or of a bacterial suspension less than 1μm.

Calculations are carried out by means of an electronic laser counter. Figure 5.3 shows (β depending on the diameter of the particle) the result obtained with a rest filter. Depending on the industry, the desired β index can be:

$$\frac{5000}{1} \text{ or } \frac{2000}{1} \text{ or even } \frac{200}{1}$$

It should be emphasized that passing from a nominal removal threshold to an absolute removal threshold may raise the cost by several dozen times, and it is not always worth it.

- **Fouling Index FI :** The lowering of this index relates to a reduction of the fouling potential of water; it is often far more significant (in colloidal water) than the size of the particles alone, which are theoretically removed.
- Parameters associated with use
- Regeneration or consumption of support material.
- Head loss or admissible filter run.
- Suspended solids concentration in the water to be created.
- Risks of releases: supports or retained matter.

Types of Filters

Differentiating between a thin support and a thick support becomes deceptive in view of the removal thresholds currently desired. Different ones are:

- **Expendable cartridges** which are equipped with:
  - pleated membranes made of paper, polycarbonate or Nylon 66 films, unwoven, heat-welded fabrics (polypropylene), which are usually not subject to release. They may have absolute removal thresholds between 0.1 and 20μm, and can be distinguished by their filtering areas,
  - felt, unwoven or wound textiles, plastic composites, which can be released.
  - Cartridges backwashable with water, equipped with:

- fibres and sintered metals with nominal thresholds ranging from 6 to 100 µm,
- fabrics made of monofibre horsehair (polyester) with nominal thresholds of between 20 and 100µm.

Backwashing is only possible with cartridges with a high nominal threshold used in water with a low SS content.

- **Regenerable candles** equipped with:
  - sintered metals or ceramics,
  - agglomerated plastic materials.

Regeneration is less often ensured by backwashing using filtered water than by other methods that are especially suited for, and compatible with, the supports (steam, acids, ultrasounds, etc.).

Generally speaking, whether it concerns backwashable cartridges or regenerable candles, the increased fineness of the filtering supports and the filtration of clogging water result in a progressive deterioration of the supports; they must be replaced after a certain number of cycles.

Filtration on Precoated Support

This is in-depth filtration through a transitory media maintained by a thin support and formed either by the introduction of an exterior precoat (a filter aid) or by the slurry to be filtered itself.

In the initial phase of the operating cycle, precoating takes place, which consists in recirculating the filter aid or the slurry itself on the filter in such a way that the coarsest particles can mix and form arches over the apertures of the thin support, the openings of which are several times larger than the diameters of the particles. Thus a filtering precoat can form and build a filter cake, the thickness of which determines the filter run between washings.

During the filtration cycle, it may be necessary to continuously inject a filter aid, either to slow down the increase in head loss or to improve the filtrate. This is called feeding.

- **Applications**

Filtration through a precoat is used for liquids that are usually only slightly loaded with SS, and filter runs progress from several days to several weeks:

- power station condensates,
- oily condensates from heating,
- beers and wines,
- aqueous cutting fluids from truing process,
- hydraulic fluids,
- syrups from sugar and glucose mills.

Filtration with a self-forming precoat is used with slurries containing 0.5 to several $g.l^{-1}$ SS, and a filter run can be limited to a few hours. For instance:

- cloudy juice from carbonation in sugar refining;
- wort from brewing operations, settled product in vats,
- pulps from hydrometallurgy.

- **Precoat and feed materials**

Depending on the application, different materials can be used:

- **cellulose**, in the form of high-purity fibres, has a filtering capacity comparable to that of a slow filter paper, but it has a very low adsorbent capacity. It is insoluble cold or warm water, and starts to hydrolyse at 85°C.

- **diatomaceous** earth consists of fossilized siliceous shells of marine origin; it is very fine (5 to 100 μm) and has some adsorbent capacity. In the presence of water loaded with colloids, it provides better clarification than cellulose.

  It is also able to adsorb emulsified impurities such as oils or hydrocarbons. The silica of the diatomaceous earth is slightly soluble in demineralized water (especially when alkaline),

- activated carbon, because of its very high adsorbent capacity, can be used on a supporting layer of cellulose or diatomaceous earth for colour removal and for almost complete removal of organic matter of vegetable origin,

- ion exchange cation and anion resins in powder form, mixed in varying proportions, provide filtration with thorough removal of colloidal iron or total demineralization of condensates in thermal and nuclear power stations.

- **Washing**

Backwashing is carried our when the design permissible head loss reaches the maximum. In order for washing to be effective, the precoat and the suspended solids retained on the precoat must be completely detached from the support without causing progressive clogging.

FILTRATION ON A GRANULAR BED

Principle and Monitoring

Water to be treated passes through a filter bed made of a granular material; the layer depth is an important parameter that depends on the type of filter used. Suspended solids are retained in the intergranular spaces throughout the greater part of the layer depth. The operation of the filter is usually monitored in two ways.

Measurement and Evolution of the Quality of the Filtered Water

Figure 5.4 shows the evolution of turbidity in a filtrate and defines the typical periods of a filter's operation:

c = maturing period,

b = period of normal operation,

d = initiation of filter breakthrough,

e = acceptable limit of turbidity; the turbidity of filtered water reaches this value for a time t_1.

Measurement and Evolution of Head Loss

The graph in figure 5.4 shows the variation of head loss P with time. The design of the unit allows for a maximum head loss that the filter should not exceed; for example, P_2 = 150hPa (=150 cm WC). This head loss is reached after a time t_2, which is an important characteristic of filter operation.

Optimizing Filter Operation

In order to obtain optimum operation of the filter, it is important that the filter attain a head loss P_2 corresponding to time (t_2) before breakthrough at time (t_1), which is:

$$t_1 > t_2$$

For water coagulated with a metallic salt, t_1 and t_2 are given by the following test formulae (Richard and Croce-Spinelli), which indicate the variation of t_1 and t_2 as a function of operating characteristics:

$$t_1 = a.v^{-0.95}.K^{0.75}.D^{-0.45}.L^{0.95}.V^{-1.85}$$

$$t_1 = b.v^{-0.75}.K^{-0.7}.D^{-1.5}.P^{0.9}.V^{-0.65}$$

with:

D: effective size of the material,

L: layer depth,

P: rise in head loss,

V: filtration rate,

K: cohesion of retained floc,

v: volume of flocculated suspended solids in the water to be treated.

The effective size of the filtering material and the cohesion of the retained floc are essential factors in the variation of t_1 and t_2. The coefficients a and b are experimental. After a single filtration test, the formulae make it possible to predict the different rimes t_1 and t_2 corresponding to various operating conditions.

Pressure Curves

The graphs in Figure 5.5 represent an open filter, with a sand depth BD and a water depth AB. On the right hand graph, the levels of the pressure take-offs A, B, C, D as measured from the floor D of the filter are plotted on the Y-axis, and the pressures represented as water head are plotted on the X-axis with the same scale as on the Y-axis. Thus at point B of the filter, at the top of the filter bed, the pressure is always equal to the water depth AB, plotted as B'b. At point C of the filter bed, when the filter is shut down, the pressure takes the value AC, plotted as C'_{Co}. Likewise, the static pressure at floor level equals AD, plotted as D'_{do}.

All the points representing the static pressure at different levels of the filter are on the 45° straight line A'_{do}.

With the filter in operation, according to Darcy's law, the head loss in homogeneous; clean sand is proportional to the depth of the sand and to the flow rate, which is taken as constant for this analysis. The pressure at point C of the filter becomes equal to C'_{C1}, with the value C_oC_1 representing the head loss of the sand between levels B and C; likewise, at floor level, the pressure at D becomes equal to D'_{d1}, with

d_0d_1 as the head loss in clean sand.

The line bc_1d_1 is a straight line since C_0C_1 and d_0d_1 are proportional to the depth of sand (Darcy's law).

When the sand is completely matured, the plotting of the pressures C'_{C2} and D'_{d2} at the various levels of the sand gives the curve bc_2d_2, which represents the pressures in the filter; it has a curvilinear section and a linear section parallel to the straight line bd_1 which represents the head loss with a clean filter.

Point c_2 that shows the start of the linear head loss, indicates the level C reached by the impurities in the sand; below C, when the head loss is linear, the sand is clean. Point c_2 defines the depth BC of the "filtration front" at the time considered.

Therefore, the minimum sand depth and minimum head loss anticipated before clogging are BC and c_0c_2 respectively.

The shifting of point c_2 during clogging represents the progress of the filtration front. In Figure 5.5, where the filter no longer gives clear water once the maximum head loss P_2 is reached, the curve representing the pressures at different points in the filter is given by $bc_fd_fe_f$, it reaches the floor without having a straight section, which means that the filtration front has passed the floor and filter break-through has occurred.

If a filter with a greater depth of sand had been used, the curve representing the pressure at the different points of the filter for the maximum available head loss would have become linear at point e_f; this immediately gives the minimum depth DE of sand that should have been added to make $t_i = t_2$.

Finally, experience shows that the values of ti that correspond to different depths of a specific sand are fairly proportional to the corresponding thicknesses.

Maximum Removal Capacity of a Filter

Suspended solids lodge between the grains of the filter material. Given the fact that sufficient space must always be left for the water to flow, on the average the sludge can hardly fill more than a quarter of the total volume of voids in the material.

For a 1m layer depth and 1m^2 filtering area, that is, a volume of 1m^3 of material, there are approximately 450 litres of empty spaces whatever the particle size; the volume available for the removal of particles is about 110 litres, provided that the effective size of the filter media and the head loss anticipated by the design of the unit are suited to the nature of these particles.

When the filters operate by gravity (open filter), and the suspended solids to be retained have a hydroxide floc base, their DS content does not exceed 10 g.l^{-1}, the quantity that can be removed per m^3 of filter material is therefore no more than $110 \times 10 = 1100$g.

This figure increases when the floe contains dense mineral matter (clays, calcium carbonate). For a sludge containing 60 g.l^{-1} DS, it can reach:

$$110 \times 60 = 6600 \text{ g}$$

In the case of pressure filtration of industrial impurities, the layer depth may reach 2 m and the head loss, 0.5 bar or even 2 bar. Thus, the filter can retain a quantity of matter as high as:

- $CaCO_3$: 4 to 15 kg per m^2 of filtering area,
- oily slime: 10 to 25 kg,
- mill scale: 20 to 100 kg.

These values indicate the maximum permissible content of suspended solids in raw water entering a filter once its filtration rate and the run between two washing operations have been determined.

For example, a filter with aim deep bed operating at a rate of 10 $m.h^{-1}$ that requires washing every eight hours (80 m^3 water per m^3 filter bed between washing operations) cannot cope with more than:

$\frac{1100}{80}$ = 13.75$mg.l^{-1}$ flocculated suspended solids, or more than

$\frac{6600}{80}$ = 82.5 $mg.l^{-1}$ dense mineral suspended solids.

For suspended solids in river water, the figure will be midway between the above two values.

The Porous Media

Physical Properties

A filtering material is generally defined by the various factors.

- grain size,
- effective size (ES),
- uniformity coefficient (UC),
- **grain shape:** angular (crushed material) or round (river and sea sand). The same filtered water quality is obtained using an angular material whose effective size is smaller than that of a round grain material.

  Given an equal size, the head loss increase is less with coarse grains than with round grains for, contrary to what might be expected, coarse grains bed down less easily than round grains, and leave larger spaces for the water to pass through,
- **friability:** it allows suitable filter materials to be selected without the risk that the washing operations will produce fines.

  A material that is too friable is unacceptable, especially with downflow filters where the washing ends with an expansion phase with water only, as the fines formed clog the filter surface,

- **loss in acid:** obviously, a high loss in acid cannot be tolerated when the water is likely to contain corrosive carbon dioxide gas or any mineral acidity,
- the density of the grains making up the filter media,
- their bulk densities in air and water.

There are other properties specific to adsorbent materials such as activated carbon;

Nature of the Porous Media

Quartz **sand** was the first material that was used for filtration, and it is still the basic material in many existing filters.

Anthracite or **marble** can be used instead when any trace of silica must be avoided in industrial processes or when they are easier to obtain.

For some methods of treatment, such as polishing, tertiary treatment of effluents, etc., it is advantageous to use materials with a large specific area, *e.g.*, expanded schists, Biolite, pozzuolana or other similar material.

Some filters use a combination of different materials (multi-media filters). In this case, the sand may be combined with anthracite, garnet, schists of varying porosity, etc., provided that these materials have low friability and low loss in acid.

Finally, filtration may be effected through sufficiently strong granular activated carbon in the following cases:

- to replace sand after settling treatment both to remove the residual floe and to combat pollution by adsorption,

- in a second filtration stage for polishing treatment only or dechlorination.

Choice of Grain Size for a Filter Media

This choice must be made while also taking the depth of the layer and the filtration rate into account. It depends on the nature of the water to be filtered (direct filtration of raw water, filtration of settled water, biological filtration of secondary or tertiary wastewater), and on the desired quality of water. It also depends on the type of filter used (pressure filter or gravity filter) and on the available head loss. The following table shows the influence of various parameters on the quality of water and the filter runs.

	Diameter of the grains	Layer depth	Filtration rate	Available head loss
Quality of filtered water				=
Filter run				
Loading per m^2	=		=	

The direction of filtration is generally downward. Depending on the type of washing system selected, there are three types of filtration which correspond to a choice of different grain sizes:

- **filtration on a layer of homogeneous material:** This material is washed with air and water without hydraulic expansion during the final rinsing phase. This results in a perfect homogeneity of the filtering layer; the size of the grains in the filter material is the same at the bottom and the top of the filtering layer.

- **filtration on a layer of heterogeneous material:** When the washing process uses water alone or when there is a final rinsing phase with hydraulic expansion of the material, classification of filter material occurs. Here, the coarsest grains are at the bottom of the filter while the finest are at the surface. During the filtration cycle, the fine filtering material receives the water to be purified which contains all the matter to be retained, while the coarsest filter material receives a cleaner water. Thus, controlling the filtration cycle is more difficult; the cycles are shorter given the fact that the fine material screens the water at the surface of the filter.

- **filtration through a multi-media filter:** the observations mentioned above have led to the setting up of a filtration process using two filter layers (dual-media filtration) and even several filtering layers (multi-media filtration).

To avoid the screening effect resulting from the finest grains of sand in filtration on a heterogeneous layer and to promote the penetration of impurities throughout the entire depth of the filter, part of the fine sand must be replaced by a layer of lighter material with an effective grain size greater than that of the sand. The grain size for each of the two layers must be carefully selected for it enables them to undergo similar expansion with the same flow of wash water, thus enabling them to be reclassified at the end of each washing, prior to resuming the filtration cycle.

This rule (which dates back to 1880 when it was used by Smith, Cuchet and Monfort), allows distribution of the retained SS: the coarsest are retained in the upper layer which is composed of large grains, while the lower, fine-grained layer performs a polishing and safety process.

There are also filters that consist of three or more media; they improve the in-depth penetration of the impurities, although their use imposes a variety of conditions affecting the choice of materials and the washing technique employed.

Washing the Filtering Media.

Washing is an extremely important operation, which, if inadequately done, leads to permanent clogging of some areas resulting in only a small passage for the water. The head loss increases more rapidly, and filtration is locally faster and less effective.

The filtering material is washed by a current of water flowing from the bottom upwards, in order to dislodge the impurities and convey them to a discharge channel. At the same time the filtering material needs to be agitated in the current of water. A number of methods can be employed to achieve this result.

Washing with Water alone to Expand the Filter Bed

The current of water must be sufficient to expand the filtering material, *i.e.*, to bring about an increase in its apparent volume of at least 15%.

As the viscosity of water varies according to temperature, it is desirable that a system should be provided for measuring and regulating the flow of wash water so as to keep the degree of expansion desired constant over time.

The expanded layer then becomes subject to convection currents; in certain zones the filtering material moves downwards and in other neighbouring zones upwards.

Because of this, portions of the compact layer of sludge encrusting the surface of the filtering material are carried deep down to form hard and bulky mud balls as a result of the action of eddy currents. This is partly overcome by breaking up the surface crust with powerful jets of high-pressure water ejected from fixed or rotating nozzles (surface washers).

This method requires considerable care and makes it necessary to measure the expansion of the filtering material exactly. Its greatest drawback is that it results in a size grading whereby the finest filtering material is concentrated on the surface; it is therefore an unsatisfactory method for downward filtration.

Simultaneous Air and Water Washing without Expansion

A second method, which is now widespread, is to use a low backwash flow rate which will not cause expansion of the sand, and at the same time to stir the sand by an injection of pressurized air. Thus, the sand remains stable, and the surface crust is completely broken up by the air; in this way, mud balls cannot be formed; in fact, they do not occur with this type of washing process.

During this period of **air scour**, the higher the flow rate of the wash water, the more rapid and effective the washing will be. The minimum value for a washing to be effective and the maximum value not to be exceeded so as to avoid a loss of filtering material both depend on the material and on the filter parameters.

When the impurities have been removed from the filtering material and collected in the layer of water between the sand and the discharge channel, "rinsing" must take place, *i.e.*, the layer of dirty water must be replaced by clear water.

Rinsing may be carried out by various methods after air scour has Stopped, such as the following:

- continuing the backwash at a constant flow rare until the discharged water runs clear. The time this takes is inversely proportional to the flow rare of water (which must always be higher than 12 $m^3/h.m^2$), and proportional to the depth of the layer of water above the filtering material,
- increasing the flow rare of water during rinsing to at least 15 $m^3/h.m^2$,
- sweeping the surface of the filter with a horizontal current of raw or settled water combined with the backwash,

- draining off the dirty water above the sand and sweeping the filtering material surface as above.

Washing with Air and Water in Succession

This method of washing is used when the filtering material is of such a nature that it is impossible to use air and water simultaneously without running the risk that the wash water will carry off the filter media to the drain. This applies to filter beds composed of fine sand or low-density materials (anthracite, activated carbon or Biolite, etc.). This type of washing is also used for dual-media filter beds.

In the first stage of the washing operation, air is used alone to detach the retained impurities from the filtering material. In the second stage, a backwash of water with a sufficient velocity to bring about the expansion of the filtering materials) enables the impurities detached during the first stage to be removed from the bed and to be carried away. In the case of impurities which are heavy or particularly difficult to remove (for example wastewater), this cycle may be repeated several times.

Washing by Sections

Usually, washing involves the entire surface of a filtering unit. In some types of filters it may be done by sections.

In the filter, the fixed walls mark off the cell units for washing. An apparatus moves into place above each of the cells at a time. The dirty water is drawn up through the sand layer of the section and the water is sent into a side discharge channel. The water to wash a cell comes directly from neighbouring cells. The cells are washed one after the other.

This type of washing may be continuous. Its major drawbacks are:

- the washing takes place with water alone, *i.e.*, without air,
- it is impossible to limit the flow of water filtering through the cell that resumes filtration; thus, the velocity is higher because there is no cellular regulating system and, in certain cases, this can result in a deterioration of the quality of filtered water,
- there is a risk of cell isolation being insufficient.

This type of washing can be justified in certain cases (filtration of cooling system water, of wastewater, etc.). However, it is not safe for the filtration of good quality water (drinking water, etc.).

Frequency of washing and wash water requirements

The frequency of washing depends on the nature of the water to be filtered as well as on the nature and the quantity of the SS to be retained. Washing must begin as soon as the head loss reaches its maximum value or when filter break- through occurs. In practice, the washing operation is often carried out after a certain volume of water has been filtered, according to operating conditions and usage experience.

The amount of wash water consumed depends essentially on the character and weight of the parades retained per m^3 of filtering material. The combined use of air scour and settled water makes it possible to reduce water requirements by some 20 to 30% as compared with washing with water alone.

Wash water requirements are greater:

- the deeper the layer of water above the filtering material,

- the lower the flow rate of the backwash water alone,
- the greater the distance separating the sludge discharge channels,
- the larger the quantity of sludge to be removed, .and
- the greater the cohesion and density of the sludge.

Water requirements are also increased by surface washers.

DEEP-BED FILTERS

The process of filtering water through beds of granular media in order to purify it is in general use throughout the world. Many different types of filter are used with the principal objective of removing microscopic suspended particles from water. The filters may be broadly classified as 'rapid' or 'slow' according to the rate at which they operate, with the further distinction that in slow sand filtration there is biological activity, whereas in rapid filtration physical removal is the important factor. The process is a dynamic one in which the change in concentration of the suspension flowing through the bed is a function of depth and time.

The two types of filter in common use are known as 'slow' and 'rapid'. They differ both in the rate at which they filter the influent water and in their mode of action. Slow filters treat water at a rate of about 0.1 $m^3m^{-2}h^{-1}$ (approximately 0.029 mm s^{-1}) and rely to a great extent on the formation of a surface biological layer or 'Schmutzdecke' for their cleansing action. In rapid filters the rate of treatment is typically in the range 5-6$m^3m^{-2}h^{-1}$ and the full depth contributes to purification and consequently has to be cleaned. Natural silica sand is generally used for both filter types but other media are possible, for example crushed anthracite.

Rapid filters operate at a much higher rate and so require frequent cleaning. This is done by backwashing the sand *in situ* using filtered water, with the aid of mechanical, air or water scouring. The high rate of flow and the need for wash water under pressure adds to the complications of a rapid filter. Sometimes the two systems are combined, a rapid filter being used as a 'roughing' filter to supplement a slow sand plant. The general features of slow and rapid filters are summarized in Table 1.

Table 1 : General features of slow and rapid filters.

Parameter	Slow sand filter	Rapid sand filter
Rate of filtration (mh^{-1})	0.1-0.2	5-10
Size of bed (m^2)	5-200	5-200
Depth of sand (m)	0.6-1.2	0.6-1.2
Sand effective size (mm)	0.15-0.3	0.5-1.0
Sand uniformity coefficient	< 3	< 1.5
Gram size distribution	unstratified	stratified by backwashing
Filter floor	designed for filtrate collection only	designed for filtrate collection only and back-wash distribution
Loss of head limit (m)	water depth on filter	water depth on filter
Duration of filter run between cleanings (d)	20-60	1-3
Penetration of suspended matter	superficial	deep
Method of cleaning	(1) scraping off surface layer and renewing at intervals (2) mechanical washing of surface layer *in situ*	back-wash with air and water or by water only at high rate
Amount of wash-water used for cleaning sand (% filtrate)	0.2-0.6	2-6
Pre-treatment of water	usually none	coagulation+ flocculation + sedimentation or flotation
Supplementary treatment	disinfection	disinfection
Relative capital cost	high	low

RAPID FILTERS

Rapid filters are most frequently used in water treatment following pretreatment by chemical coagulation/clarification; they are also used in wastewater treatment for tertiary effluent 'polishing'. Rapid filters are usually constructed as open-top, free-surface units (rapid gravity filter, RGF) and less frequently as in-line pressure units. Rapid filters typically contain about 0.6-1 m of sand overlying an underdrain system. The depth of water overlying the sand surface in RGF units is generally not less than 1.0 m.

Rapid Filter Underdrain Systems

The underdrain system in rapid filters is designed to transmit filtrate and to ensure a uniform distribution of back-wash water and, where used, air. It must also prevent loss of sand with the filtrate. Many types of underdrain system have been devised,. The most widely used system consists of a set of perforated pipe laterals surrounded by graded silica gravel layers, which shows a four-layer underdrain system with the grain size ranging from 2 mm in the uppermost layer to 50 mm in the bottom layer. The main function of the gravel layers is to distribute the back-wash upflow evenly over the bed area and prevent the penetration of the sand into the pipe lateral system. Pipe laterals are usually 75-100 mm diameter and are spaced at 150-225 mm centres, with orifices of 6-12 mm diameter at similar centres, discharging downwards. They are connected to a central pipe or channel manifold. The manifold and lateral system are hydraulically designed to ensure a uniform distribution of back-wash water over the filter area. Where air is used for filter cleaning, a separate air manifold and lateral system is provided. Alternative underdrain systems incorporating nozzles or no-fines porous concrete may partly eliminate the need for gravel layers and may also provide for air distribution. In terms of freedom from clogging and long-term performance reliability, the pipe lateral system is probably still the best available system. Casey (1992) presents an analytical procedure for the computation of flow distribution in pipe lateral/manifold systems of the type used in filter underdrain systems and also as flow distribution systems in sludge-blanket clarifiers.

REMOVAL MECHANISMS IN RAPID FILTRATION

While one might intuitively expect mechanical straining to be a major mechanism of solids separation in rapid nitration this is not the case. Indeed, such a separation mechanism would rapidly lead to a blinding of the filter surface and a consequent very rapid increase in head loss. In fact much of the paniculate matter removed in rapid filtration is very much smaller than the average filter pore size. The separation of such material results from a combination of transport processes which bring articles into contact with gram surfaces, and attractive forces which secure their attachment. The transport mechanisms may include diffusion, hydrodynamic forces, interception and sedimentation, while the attachment mechanisms involve physico-chemical forces.

Diffusion or random Brownian motion is of particular significance for very small particles but has a negligible transport influence compared with gravity and hydrodynamic forces for larger particles. Sedimentation is considered to be significant because of the very large upward-facing horizontal surface area within a filter onto which particles may settle. Removal by sedimentation is a function of particle size and particle density and the velocity distribution within the filter pores.

Inertial forces do not constitute a very significant removal mechanism in rapid filtration because of the low fluid velocities involved and the retarding effect of viscosity.

Surface attachment influences include the universal force of mass attraction (van der Waals force) and electrostic forces which depend on the sign and magnitude of the charges on the particles and on the sand grains (Coulomb force). These forces are only of significance when particles and sand grains are brought close together by one or more of the transport mechanisms noted above. The Van der Waals force decreases in proportion to the sixth power of the distance between mass centres, while the Coulomb force decreases in proportion to the second power of the distance between surfaces. Particle attachment is also promoted by the development on the sand grains of a sticky gelatinous layer of deposited material.

FILTER MEDIUM SELECTION

Silica sand is invariably used as the filter medium in rapid filters. It should be as near single size as possible, otherwise the bed becomes hydraulically stratified by back-washing with a grain size gradient from fine at the top to course at the bottom—the opposite to that required for efficient use of the full depth of the bed.

A grain size gradient of course at the top to fine at the bottom can be maintained by using media layers of progressively increasing density, *e.g.* anthracite on sand on magnetite. The density difference between these materials is sufficient to prevent intermixing during back-washing, provided the size difference is not too great. However, most filters contain sand only because of its ready availability and lower cost than materials such as anthracite and magnetite.

The filter media size distribution may be fully specified, *i.e.* its grading curve may be required to be within specified upper and lower limits, as indicated in Figure 5.12. Alternatively, the size distribution may be specified in terms of the Hazen effective size and uniformity coefficient. The effective size is the sieve size that passes 10% by weight of the material (d_{10}), the uniformity coefficient is the ratio of the sieve size that passes 60% by weight of the material to the sieve size which passes 10% of the material (d_{60}/d_{10}).

Because the smaller grain sizes are of greater significance to filter performance, the ten percentile diameter was chosen as the effective size. The ten percentile by weight corresponds approximately to the fifty percentile by number. Ideally the uniformity coefficient should be as close to unity as possible and preferably less than 1.5. Obviously it is better to set full grading curve limits than to specify an effective size and uniformity coefficient only—the latter specification identifies only two points on the grading curve.

Selection of bed depth and sand size is frequently made on the basis of local experience, an effective size of 0.6 mm and a bed depth of 0.6 m being commonly used. Pilot plant tests provide the best basis for the determination of the design values for these parameters. Pilot units may be constructed in cylindrical tubes of 100-150 mm diameter. After a 'run-in' period, the filter performance is monitored by measuring the head loss at the end of a filter run of the design duration T, while the residual suspended solids concentration (or turbidity level) is measured after a greater duration, *e.g.* 1.1 T. These measurements are carried out for various filter depths and the results. The use of a greater filter run time for the filtrate quality measurement than for the filter head loss measurement allows the latter to be used for the control of filter run duration with a built-in margin of safety in relation to filtrate quality.

BACK-WASHING PRACTICE

Conventional rapid gravity and pressure filters are operated on a batch mode basis and are taken out of production for back-washing at the end of each filter run. Back-washing practice varies. The following are commonly used methods:

(1) Air-scour at an upflow rate of about 30 mh^{-1} for 3 min followed by water upflow at a rate of 20 mh^{-1} for 5 min.

(2) Upwash with water only, at a rate of about 50 mh^{-1} for 5 min (used in the United States).

(3) Simultaneous air and water upflow for 5 min, air at an upflow rate of about 50 mh^{-1} and water at a rate of about 20 mh^{-1}, followed by water only at a rate of 20mh^{-1} for 5 min.

Experimental comparison of these methods (Patterson, 1978) has shown that method (3), *i.e.* simultaneous use of air and water, provides the most efficient cleaning action.

The back-washing operation requires a relatively large water flow for a relatively short period (5-10 min) during which an individual filter unit is being washed. This can be provided by pumping from low-level storage or the appropriate volume may be stored at high level to provide a gravity back-wash supply. The normal sequence of operations in back-washing is as follows:

(1) Inflow to the filter is shut off.

(2) The standing water level on the filter is drawn down to near bed surface level.

(3 The air and water back-wash valves are opened.

(4) After the set duration of the back-washing process, the backwash valves are closed.

(5) Filtration is restarted by opening the inflow control.

The back-wash water is collected in decanting channels set about 300 mm above the sand bed surface (sufficiently high so as not to interfere with bed expansion caused by back-wash).

While conventional filters are back-washed intermittently in the manner just described, innovative filter systems are being developed which permit continuous filtration and reduce back-wash water requirements (Boiler, 1994).

CELLULAR SUB-DIVISION OF FILTRATION AREA

The required filter area A is determined by the required filtrate output and the design filtration rate, taking into account the variations in both of these parameters throughout the year. The required water output should include an allowance for the quantity of filtered water to be used for back-washing. The filter area should allow for at least one filter cell being out of use for repair or alteration.

The total filter area A may be divided into n individual units, each of area a in accordance with the relation:

$$a = \frac{A}{(n-1)} \cdot \frac{T}{(T-t_w)}$$

where T is the filter run duration and t_w is the filter wash duration. The factor (n – 1) allows for one unit being out of commission for repair or alteration; the ratio $T/(T\text{-}t_w)$ allows for downtime for back-washing. In general, the value of n should not be less than 3 and the value of a should not be less than $10m^2$. Individual units of very large plan area are undesirable because of the correspondingly large flows required for back-washing—units of plan area greater than $200m^2$ are rare.

CONTROL OF FILTRATION

The hydraulic control of filtration involves:

(1) equal sub-division of incoming flow among the filter units in operation at any time;

(2) maintenance of a constant rate of flow through each filter for the full filter run duration; and

(3) operation of the back-washing sequence.

A filter run may be terminated when the filtrate quality has deteriorated below a set value, *e.g.* colour 10°H, turbidity 0.5 FTU or when the head loss has exceeded a set value. Head loss is the preferred control parameter because of its convenience of measurement. A design value of terminal head loss, consistent with the achievement of the set quality target and which also avoids negative head development within the filter, is chosen. The development of a negative head can be avoided if the maximum head loss is limited to a value equal to the depth of water above the bed surface.

All filter control systems incorporate an automatic valve on the filtrate line, which gradually opens during a filter run thereby compensating for the increasing filter resistance, so as to keep the filtrate discharge rate constant. A common feed channel with a lateral discharge to each filter unit controlled by a weir or orifice is commonly used to sub-divide flow equally among filter units.

MEMBRANE FILTRATION

Membrane processes can be broadly classified in two groups on the basis of the nature of the driving force employed in their operation. The more widely used membranes are those that rely on an imposed pressure gradient to force water through the membrane, while retaining particulates and, in some cases, solutes. The second type of membrane process uses an electrical potential gradient across the membrane to effect the selective migration of ions in a filtration process known as electro-dialysis.

Pressure-driven Membrane Processes

Pressure-driven membrane processes can be categorized by the molecular weight or particle size cut-off of the membrane, as summarized in Table 5.2. MF and UF membranes are considered to effect species separation by a sieving mechanism and hence they remove particles larger than their size-cutoff limits. The mode of action of both NF and RO membranes is considered to include both sieving and diffusion-controlled transport. NF and RO membranes are capable of separating both organic molecules and inorganic ions. NF and RO membranes are operated at high differential pressure (up to 100 bar) and have typical characteristic flux rates in the range 5-91m^{-2}h^{-1} bar^{-1}. UF membranes are operated at differential pressures in the range 1-10 bar. MF membranes are operated at differential pressures $\geq$ 5 bar and achieve flux rates in the range 100-2001m^{-2} h^{-1} bar^{-1}.

Table 2 : Classification of pressure-driven membrane separation processes.

Membrane process	Size cut-off range (um)	Examples of materials separated
Microfiltration (MF)	0.05-1.5	microbial cells, large colloids, small particles
Ultrafiltration (UF)	0.002-0.05	macromolecules, viruses, colloids
Nanofiltration (NF)	0.0005-0.007	viruses, humic acids, organic molecules, Ca^{2+}, Mg^{2+}
Reverse osmosis (RO)	0.0001-0.003	aqueous salts, metal ions

There has been an ongoing development in synthetic polymeric membranes since the 1960s which has now reached the stage where membrane filtration can be considered a promising alternative to conventional potable water treatment processes such as chemical coagulation/sand filtration/disinfection. Membranes are marketed in module or cartridge housings, which are supplied in ready-to-use form. They are fitted with feed, permeate and concentrate connections. A variety of module configurations are available (Rautenbach and Albrecht, 1989), including tubular, capillary, hollow fibre, spirally wound sheets and plate and frame. Since the permeate discharge per unit membrane area may be relatively low, modules are designed to maximize the membrane packing density, expressed as membrane filtration surface area per unit volume of module (m^2 m^{-3}).

The tubular module consists of a tubular membrane sitting inside a porous sleeve, which in turn fits into a perforated pressure-tight tube between 12 mm and 24 mm diameter. Tubular modules are normally operated in the cross-flow filtration mode.

Capillary modules consist of capillary tube bundles, each end of which is connected to a head plate. The internal capillary diameter is typically in the range 400-2500 μm and the packing density is of the order of 1000 m^2 m^{-3}. Capillary modules are available in both UF and MF ranges.

Like tubular modules, they are normally operated in a cross-flow filtration mode. Hollow fibre modules consist of hollow fibre bundles, usually bent in hairpin-like fashion, with the open ends cast into an epoxy resin head plate. The internal diameter of the fibres is typically in the range 10-40μm and the packing density of the order of 10 000 m^2m^{-3}. Unlike the capillary modules, filtration is from the outside inwards, the permeate being discharged through the open fibre ends. Hollow fibre modules are available in the UF, NF and RO ranges.

The spirally wound modules consist of membrane sheets wound around a permeate-collecting tubular core. The rectangular membrane sheets are arranged in 'pockets', each of which consists of a highly porous sheet (permeate-side spacer) sandwiched between two membranes, which are glued together along three edges. The fourth edge of the pocket is attached to the collecting tube. Several such pockets are spirally wound around the collecting tube with a feed-side spacer placed between the pockets to form a so-called 'element'.

In plate and frame modules, the membranes are arranged in parallel flat sheets, with porous spacers, through which the permeate is discharged. The feed solution flows through flat rectangular channels between the membranes. Packing densities are in the range 100-400m^2m^{-3}.

In general, membrane filters may be operated in either 'dead-end' or 'cross-flow' modes. In the dead-end mode of operation, the permeate is the only outflow from the module during a filter run and hence the removed particulate and solute species accumulate within the module. Cross-flow filtration, on the other hand, operates in a recirculation mode with the feed introduced tangentially to the membrane surface. The tangential flow promotes the self-cleansing of the membrane surface with an associated reduction in membrane fouling. The suspension can be recycled until the required yield is achieved.

MF processes are used to remove turbidity and chlorine-resistant pathogens such as Cryptosporidium and Giardia. Depending on their cut-off limit, MF membranes may also remove bacteria but do not offer complete removal of viruses, although substantial removal can be achieved if the virus population is largely associated with larger particles (Gallagher et al., 1995). They are typically operated at a differential pressure of less than 1 bar and at flux rates up to $150 l m^{-2} h^{-1}$.

UF is used in the food and process industries as well as finding increased application in water and wastewater treatment. UF membranes with a molecular weight cut-off of 100 000 were found to achieve a 6 logs rejection (*i.e.* below the detection limit) of seeded MS2 virus, while UF membranes with larger cut-off limits achieved partial virus removal (Jacangelo *et al.*, 1995). UF hollow fibre membranes are being used in conjunction with powdered activated carbon for the direct production of potable water from surface water (Cornu et al., 1995).

Membranes at the lower end of the UF cut-off range can be used to remove humic substances, including colour, from waters (Jensen and Thorsen, 1995).

NF and RO membranes have lower cut-off limits than UF membranes and are capable of rejecting inorganics as well as organic species.

NF rejection of inorganics is determined by the composition of the water, the rejection of bivalent ions being significantly greater than the rejection of monovalent ions. RO membranes remove the dissolved organics almost completely and also remove most of the inorganic salts. Hence, they are used for the production of potable water from brackish waters and seawater.

Membrane Fouling

Membrane fouling, with a consequent reduction in specific flux, is recognized as a major limitation of the membrane filtration process. It is caused by the gradual build-up on the membrane surface of a fouling layer, which may consist of colloidal particles, iron and/or manganese oxides, and a biofilm, or may be due to inorganic scale-formation, caused, for example, by the precipitation of sparingly soluble salts such as calcium carbonate, magnesium carbonate, calcium sulphate, silica, magnesium hydroxide, etc.

Pretreatment of the feed water by processes such as chemical coagulation, rapid gravity filtration or some form of cartridge filtration can greatly reduce paniculate fouling. Scaling can be eliminated by chemical adjustment of the feed water composition to ensure that the solubilities of the scaling precursors are not exceeded in the retentate.

Electrodialysis

Electrodialysis (ED) is a membrane filtration process in which ionic species are removed under the driving force of an elecrtric field. The water to be treated is pumped through a membrane stack which consists of alternately placed anion-permeable and cation-permeable membranes, as illustrated in Figure 5.16. Under the action of the applied electric field the anions migrate towards the anode and the cations migrate towards the cathode, resulting in the stripping of ions from alternate chambers and their concentration in the remaining chambers.

SLOW SAND FILTRATION

Slow sand filters have been used (Graham, 1988) for the purification of public water supplies since the early part of the nineteenth century. In the early days, prior to the introduction of chlorination for water disinfection, slow sand filtration provided an effective barrier against the transmission of waterborne pathogens. For example, Dublin Corporation constructed slow sand filters at Roundwood, Co. Wicklow, in the 1860s to treat impounded River Vartry water. This scheme is still in operation today, supplying about 77Ml d^{-1}.

Slow sand filtration effects a modest removal in colour and can only be used as a sole treatment process where the raw water colour is less than 15-20° H. Where the raw water silt content is high or variable, pretreatment by roughing filters is required. Because of the relatively high colour of many surface waters, slow sand filtration is not widely applicable as a sole treatment process. Worldwide, however, it is still a significant unit treatment process in modern water processes technology, albeit with rather limited application.

A typical slow sand filter unit is composed of 0.15-0.3 mm effective size sand and a uniformity coefficient < 3, with a bed thickness in the range 0.6-1 m. The sand bed is unstratified and is supported on a graded gravel layer, designed to prevent the ingress of sand into the underdrain system. The latter, which is designed to convey the filtrate to a central collector channel, may typically consist of a system of perforated lateral pipes discharging to a central manifold pipe or collector channel. Many other types of proprietary underdrain systems are also available, including perforated floors and narrow slit nozzles, the use of which may eliminate the need for the gravel layers. Very small aperture systems should be avoided as they are prone to the development of increased head loss with time due to the accretion of material on the material surface. Silica sands and gravels are used.

Biological activity plays an important role in slow sand filtration. After a few weeks operation the topmost layer of sand grains become coated with a gelatinous biological film, which forms a continuous surface mat or 'Schmutzdecke' consisting of bacteria, algae, protozoa and colloidal material derived from the water. This membrane is considered to have a significant barrier-type role in the slow sand filtration process. Bacterial action, resulting in the oxidation of biodegradable substances, extends some distance into the sand bed.

The length of filter run is typically of the order of 30 d. The head loss through the bed, which initially may be about 50-75 mm, increases as the pores in the upper layers of the filter are reduced in size by the deposited material. To avoid negative pressure development within the sand bed (see Figure 5.8) the permissible head loss is usually limited to a value equal to the depth of water above the sand surface, which is typically in the range 1.0-1.5 m. When the head loss has reached the limit value, the bed is drained down and cleaned. The usual method of cleaning is by manual skimming the top 12-20 mm of sand. The filter is then recharged slowly by the upflow of filtered water from storage.

The skimmed sand is washed and stored for subsequent re-use. Re-sanding is carried out when the skimming operations have reduced the sand bed thickness to about 0.5 m. At large slow sand filter installations the process of filter cleaning has been mechanized, thus reducing filter downtime. The London Metropolitan Water Board (Rachwal et al., 1988) uses mechanical skimming machines which incorporate an open bottomed box that penetrates 0.15 m into the sand, spanning the width of the filter. The surface is agitated and the sand is cleaned by means of high-pressure water jets which separate the deposited material from the sand grains and remove it from the filter in a washwater stream.

SUSPENDED SOLID REMOVAL BY SOLID SETTLEMENT AND FLOTATION

This chapter examines available techniques for suspended solids removal, either by settlement (sedimentation) or notation. Settlement is by far the more common technique in wastewater treatment; it involves little mechanical equipment and is the more stable to operate. There are some situations where notation is the better choice however, or where settlement might not even work:

the removal of fats and proteins from food industry effluent, or fine inorganic particulates produced during mining are classic examples.

Clarifiers generating an intermediate level sludge blanket are widely used in the water supply industry and are successfully used for fibrous solids removal in, for example, the paper industry. Simple clarifiers placed on top of a conventional settlement tank and using wedge wire mesh or gravel media, are useful for upgrading solids removal performance, or where secondary treatment is overloaded. Very high quality effluents with negligible solids content are possible after passage through upward flow clarifiers.

At the 'low-tech' end, lagoons have much to recommend them where there is plenty of space and possible odour generation is not a concern.

INITIAL CONSIDERATIONS

Solids removal dramatically reduces the 'strength' of most effluents and is widely practised. It is one of the simplest and most cost-effective ways of reducing trade effluent treatment charges.

Industrial effluents discharging to estuaries and coastal waters are frequently only treated by settlement. The sludge is often capable of reuse, and if inorganic, the liquid supernatant may be virtually unpolluted and similarly recycled as second-grade washwater.

As a general rule, effective settlement of most effluents reduces the solids by it least 50%, with a proportional reduction in oxygen demand or 'strength' measured as Biochemical Oxygen Demand (BOD) or Chemical Oxygen Demand (COD) and ranging between 30 and 50%. Tanks that are lightly hydraulically loaded and receiving effluent with coarse or well-mineralised solids may achieve 90% removal, as might a clarifier generating a sludge blanket. A 10% reduction of bacterial numbers during primary sedimentation is usually assumed in design calculations for sewage treatment.

Both settlement and flotation is easy in theory; in practice, some attention to details pays dividends in effecting satisfactory separation of solids and water.

The density difference between most waterborne solids that will form sludges and water is marginal; most organic solids lie in the range 1.03—1.10. Any major disturbance to the system-changing flow rates, turbulence, thermal stratification or sudden withdrawal of sludge or liquid—that upsets quiescent conditions within the tank, affects settlement.

Settlement tanks can have a useful balancing action in themselves by buffering sudden changes of organic and hydraulic load applied to secondary treatment, and, if carefully sized, iron out flow variations through the rest of the plant. To effect maximum settlement, however, some form of flow balancing before settlement tanks may be necessary, particularly in industrial environments where 'wash-ups' occur at the end of each shift or production batch.

Addition of chemicals to assist settlement by coagulating particles or chemical precipitation is essential for treating a variety of industrial wastewaters particularly those containing metals.

The Victorian method of adding lime as a slurry to sewage has been revitalised recently and is worthy of mention as it can be a useful physicochemical treatment for an overloaded works, reduce bacterial numbers in discharges to sea where the high pH will be of less concern and assist in phosphate removal prior to discharge to inland waters where eutrophic conditions exist.

The use of both lime and aluminium sulphate in an upward flow clarifier are described by Jones *et al.* (1991). The increased sludge volume produced by such treatment

(25%—30% is typical) is a major operational and cost consideration.

To summarise, sedimentation in some form is thus essential for most systems, and must be operated effectively to prevent deleterious effects on later treatment stages. A number of industrial effluents can be successfully treated to acceptable discharge standards by well designed settlement facilities with regular desludging, preventing the need for further and more elaborate secondary treatment. This possibility should always be examined in any review of treatment options.

ASPECTS OF SEDIMENTATION SPECIFIC TO SEWAGE TREATMENT

Sedimentation is the first and primary treatment stage of conventional sewage treatment for biofilter and activated sludge systems, although some extended aeration plant dispense with any form of primary treatment and treat crude sewage after screening and grit removal. The resultant raw sludge has a strong and unpleasant odour, requires some form of treatment before disposal, and is typically 4%-5% dry matter (DM).

Primary tanks are normally desludged at intervals of between 8 and 24 hours, which allows some consolidation to occur and produces a thicker sludge, whilst avoiding septicity and anaerobic conditions which might cause odour problems and rising sludge. Individual circumstances will apply to each site, and are best determined by local knowledge or past experience as to the 'freshness' of the sewage arriving for treatment.

Many European sewage discharges to estuaries and the sea are at present treated only by primary sedimentation, and the resultant settled sewage discharge may be acceptable in achieving quality and bacterial standards with adequate dilution and outfall placement.

Primary Sedimentation

The term 'primary sedimentation' in wastewater treatment technology refers to the separation of settleable solids from raw wastewater. The settleable solids in municipal sewage and many organic industrial wastewaters include flocculent and discrete-settling particles. Primary sedimentation tanks at municipal works have to cope with considerable variations in loading, especially when serving combined sewer systems. Primary tanks are typically designed for a maximum hydraulic loading of 3 DWF (dry weather flow). The diurnal flow variation in dry weather is generally in the range 0.5-1.5 DWF. Because of the flocculent nature of the raw sewage solids, solids removal is influenced by retention time as well as by surface loading.

The design surface loading for primary sedimentation tanks is typically in the range 1.0-1.5 m h^{-1} at maximum flow. However, it would appear that significantly higher overflow rates could be used without incurring an excessive penalty in performance, as may be seen from the pilot plant data summarized in Table 3. The liquid depth is usually in the range 3.0-4.0 m, giving a retention time of 2-4 h at maximum flow.

Table 3: Primary sedimentation pilot plant test results.

Surface loading rate ($m^3\ m^{-2}\ d^{-1}$)	Mean influent Sewage solids (mgl^{-1})	Sewage solids removal efficiency (%)
25	411	49
50	402	43
100	355	36
150	365	34

At small plants, upward flow tanks of hopper-bottom construction, are used. Tanks of this type allow sludge thickening and removal without the aid of mechanical equipment. Because of their shape, however, they are expensive to construct. Rectangular tanks are also used for primary sedimentation, but usually only at large works. The length : width ratio is typically 3-4:1, while the length : depth ratio should preferably not exceed 10 (to avoid scouring effects). Travelling bridge scrapers or submerged flight-type scrapers are used to move settled sludge to hoppers at the inlet end of rectangular tanks. The floor slope (towards the inlet end) is generally about 1:100.

Primary sludge is preferably withdrawn continuously from sedimentation tanks so as to avoid septicity and the generation of biogas which reduces the effective density of the settled solids and thus disrupts sedimentation. Primary sludge has good thickening characteristics and can be concentrated in primary tank hoppers to a solids content of 4-6%. The term 'secondary sedimentation' refers to sedimentation processes following biological (activated sludge, biofiltration) or physicochemical secondary treatment processes. The performance target for secondary sedimentation tanks is generally expressed in terms of a limiting effluent solids concentration, typically $\leq$ 30 mgl^{-1}. The sedimentation basin configurations used in secondary sedimentation applications are generally the same as described for primary sedimentation.

Factors other than those already discussed require to be taken into account in the design of secondary sedimentation basins associated with activated sludge processes. These include the basin solids flux capacity and sludge recycle rate. The activated sludge concentration in aeration basins is typically in the range 2000-4000 mgl^{-1} and thus is significantly higher than other suspensions encountered in wastewater treatment.

The concentration of suspended solids (humus sludge) in biofilter effluents is characteristically variable but is usually less than 100 mgl^{-1} Secondary sedimentation basins following biofilters are generally sized on the basis of a design surface loading in the range 1-1.5 m h^{-1}

The metal hydroxide/organic sludges resulting from physicochemical treatment are flocculent in nature and have settling characteristics generally similar to flocculent activated sludge. They can be separated by conventional sedimentation processes at surface loading rates in the range 0.8-1.2 m h^{-1}.

SEDIMENTATION IN WATER TREATMENT

Plain sedimentation of source water is rarely necessary in the production of potable water since the settleable solids concentrations in surface waters used as drinking water sources are invariably very low. However, where, river sources intermittently carry large silt loads under flood conditions, it may be necessary to have either intermediate reservoir storage or primary sedimentation as part of the overall treatment system.

Natural surface waters generally contain suspended and colloidal material resulting in colour and turbidity levels not acceptable in drinking water. Dissolved/suspended colour and colloidal solids are converted to a settleable form by the process of chemical coagulation. The resulting flocculent suspensions are clarified by sedimentation and filtration processes.

Sedimentation tanks of the kind already described for wastewaters may be used for this purpose. However, upward flow fluidized beds or so-called 'sludge-blanket' clarifiers are more usually used for this application.

Sludge-Blanket Clarifiers

Sludge-blanket clarifiers combine the roles of flocculation and floe separation. The sludge blanket, which has zone-settling characteristics, is fluidized by the upward-flowing water. Sludge is withdrawn from the top blanket, where there is a well-defined interface between the blanket and the supernatant water. Inflow is at floor level, where there is considerable turbulence and mixing which helps to promote flocculation. The earlier form of the sludge-blanket clarifier was of the hopper-bottom type. The hopper-bottom shape has the advantages of a simple inlet system and of providing a simple means of concentrating settled sludge to permit easy removal. Hopper-bottom tanks are, however, expensive to construct; they are limited in plan area if excessive depth is to be avoided, requiring multiple units in large plants. The hydraulic characteristics of these units are also unfavourable owing to the non-uniformity of the divergent upward flow, which causes short-circuiting and blanket instability. For these reasons the surface loading of hopper- bottom sludge-blanket clarifiers is usually limited to 1.0-1.5 m h^{-1}. This range can be extended upwards somewhat by the use of special energy-dissipating inlet devices.

The flat-bottom type of sludge-blanket clarifier typically has a pipe manifold inlet system, located near floor level and discharging down wards. In some designs the flow is pulsed to improve flocculation and prevent the deposition of solids on the tank floor. Clarified effluent is withdrawn by a system of decanting channels at the water surface. To avoid disturbance of the blanket surface by the upward How, the spacing of the decanting channels is usually not greater than twice the distance between the top of the blanket and the water surface. The design surface loading for flat-bottom sludge-blanket clarifiers is normally in the range 2-4m h^{-1}, depending on the settling characteristics of the floc.

Improving on the Sludge Blanket

The sludge blanket principle can be carried further, but detailed ways of doing so depend on the application. The main two groups are industrial and municipal designs. As a very crude generalization, municipal plant is large, conservatively sized and (whether from prejudice or to suit criteria in adjudicating capital expenditure) has a minimum of moving machinery, while industrial plant usually treats a smaller flow and is built with an eye to low capital investment.

Industrial sedimentation tanks mostly improve on the simple hopper bottom design with sludge recirculation, using paddles or impellers to take old sludge and contact it with the incoming water. On the very large scale these paddles could give trouble with excessive tip speeds unless the design uses a multiplicity of units, and altogether this kind of design fits better into the industrial sector. There are a number of proprietary designs on the market which vary in detail rather than in principle.

Many municipal schemes are still being designed with simple sludge blanket tanks but one interesting design is worth singling out. Its manufacturer describes a tendency for sludge blankets to form cracks which present a preferred path for water and floe to slip through. The blanket is consolidated by imposing a pulse on the incoming flow by applying an intermittent vacuum on a chamber in the inlet line. The flow rises to peaks of five times the mean flow which consolidate the blanket. This device (which has no large mechanical moving parts) can sometimes also be used to upgrade an existing tank.

THE PRINCIPLES OF SETTLEMENT

The basic theory is simple. Effluent enters a tank, either in batches or continuously, where a combination of low flow rate and time allows suspended solids particles to descend by gravity and form an underlying sludge. The clarified top liquid layer, the supernatant, overflows or is decanted as effluent for discharge or further treatment.

Particle settlement rate is dependent on three main factors:

(1) particle size;

(2) the density difference between solid and liquid phases;

(3) the liquid viscosity.

The relationship is expressed mathematically in Stokes' equation:

$$U = \frac{gD^2(d_1 - d_2)}{18u}$$

Where:

U is the terminal velocity of the particle.

g is the Waste Water Treatment by Biological Processes due to gravity.

D is the particle diameter.

d_1 is the particle density.

d_2 is the fluid density.

u is the fluid viscosity.

The settlement rate thus varies directly with the square of the particle diameter and, if the fluid is water and d_2=1, with the particle density.

Stokes' Law applies to regular, spherical discrete particles, not shapes associated with most sludge solids and particularly fibres where settlement rates will be significantly less than theoretical. However, the relationship with the square of the diameter provides the basis for encouraging flocculation and agglomeration of particles. Addition of chemicals such as polyelectrolytes, which reduce the electrostatic forces of repulsion between suspended matter, are added to assist flocculation, but are not essential.

Slow mechanical stirring and quiescent conditions will encourage flocculation by particle collision and attachment, and are just as effective with biological and highly organic sludges. Similarly, Stokes' Law accounts for the superior settling characteristics of more dense, granular primary sludges by relating particle density to settlement rate.

SETTLEMENT TANK DESIGN PARAMETERS

There are three basic considerations.

Upward Flow Rate

As solid particles are gravitating downwards against an upward flow of liquor, the upward flow rate (UFR) is of prime importance. This is expressed in metres per hour, and is also known as the surface loading, or surface overflow rate, defined by the equation:

$$\frac{\text{Effluent feed rate/hour}(m^3/h)}{\text{Tank surface area}(m^2)}$$

The water industry designs conservatively for this parameter and tanks are sized to give UFRs of 0.5—1.0$m^3/m^2/h$, which will cope with the lightest of activated sludges; 1.2m/h is an often-quoted design parameter for 'average' conditions. Heavier industrial solids may well settle out at flow rates above 2$m^3/m^2/h$, but sudden flow variations are likely to cause settlement problems at high UFRs resulting in less than 50% solids removal, which is the performance level to aim for. An effluent high in dissolved solids or soluble BOD is unlikely to settle out effectively at the higher rates unless chemically dosed before entering the settlement stage.

A simple way of increasing the weir surface area and effective length — and this is widely practised in sewage treatment – is to fit V-notch weirs around the tank. As the flow rate increases, the liquid level rises further up the V-notch and so has a larger area to flow over. Tanks which suffer a wide flow variation can benefit settlement by fitting a number of such weirs shows an installation where two are fitted, the effluent flowing into a separate channel suspended within the tank.

Retention Time

The second design consideration is to give sufficient time for the solid particles to reach the tank base or underlying sludge layer, *i.e.* the retention or detention time of the effluent within the tank. This is simply defined as:

$$\frac{\text{Tank volume}(m^3)}{\text{Effluent feed rate}(m^2/h)}$$

When sizing settlement tanks, the retention time at the maximum likely flow rate for an anticipated 50% + solids removal is the important calculation. A realistic appraisal of flow variations is essential, as there may be flow rate changes of ten times in some industrial situations, or where the effluent is entirely pumped to the tanks. In practice, retention times vary quite widely in wastewater treatment. As little as 2 hours is possible even with organic raw sludges and particularly heavy inorganic solids, but it is normally regarded as a practical minimum. Five to eight hours is a typical range in most primary sedimentation tanks at sewage works, the primary tanks of small works often being desludged at intervals longer than 24 hours without any problems, providing the sewage is not septic on arrival. Secondary settlement tanks are normally sized for a retention time of 4-5 hours, allowing sufficient time for the sludge to settle but for little consolidation other than some agglomeration to occur, so avoiding denitrification and floating material.

Some industrial operators will have tanks where the retention time is measured in days, and providing the sludge is inorganic or does not decompose or react readily, this will allow compaction to occur and give a useful reduction in sludge volume. Related issues dictating against very long retention times include temperature, both ambient and that of the effluent, with the attendant risk of decomposition, and the thickness or dry matter content of the settled sludge that can be handled easily. Most sludges will not How readily out of valves or down pipes if more than 10% DM, and pumping costs can be high at these levels.

Long retention times and sludge compaction may therefore bring greater operational problems than a regular desludging regime which produces a greater volume of thinner sludge.

Maximum Solids Loading (MSL)

The third design parameter relates to the likely suspended solids levels in the effluent entering the tank, from which it is possible to calculate the MSL, *i.e.* the mass of solids per unit area per unit time that can be applied to that tank before solids flow upwards and over the weirs with the effluent.

This parameter has been extensively researched in relation to the design of secondary settlement tanks on activated sludge plants, but applies equally to settlement of metal hydroxide sludges. WRC (1975) describes how this parameter can be measured, and is essential reading for detailed study. In summary, the solids loading is made up of two components: that due to the settling of sludge floes under gravity, and the sludge withdrawal rate from the tank. The MSL can be obtained by plotting solids loading in $kg/m^2/h$ against solids concentration in g/l. Relationship between solids loading and solids concentration below the critical rate of sludge withdrawal.

Cf is the feed solids concentration (3.5g/*l*).

Cu is the sludge withdrawn concentration in g/*l*.

The lowest point in the curve C_1 represents the maximum solids loading that can be applied before the solids rise and leave with the effluent, and also the critical value for sludge withdrawal. Under normal operating conditions, therefore, the rate of sludge withdrawal determines the solids loading that can be applied and should be above the critical value.

The settling characteristics of the sludge are equally important and can be measured as follows. A sample of effluent is slowly stirred in a cylinder of about 41 capacity, and after 0.5h the volume of settled sludge in milliliters is noted. This is divided by the solids concentration to give the stirred specific volume (SSV) in ml/g. By convention, the SSV is determined at a solids concentration of 3.5 g/l, this being a typical suspended solids value of influent activated sludge in secondary settlement tanks.

The SSV, sludge withdrawal rate, solids loading and flow rate through the tank are all related by a nomograph. In practice, maximum solids loadings on activated sludge plants are about 20% lower than those predicted by the nomograph, and some conservative sizing of tanks is therefore wise. SSV values that might be obtained are dependent on the sludge age in the plant, but 80—140ml/g at 3.5g/l concentration is a reasonable range, and anything over 150ml/g, particularly in an extended aeration plant, should be carefully monitored.

SETTLEMENT TANKS AND PRACTICAL OPERATION

Lagoons

The simplest form of settlement tank need consist of nothing more than an excavated area with earth walls. If an industrial site has the luxury of spare or derelict land, this option should be given more than a cursory glance in assessing treatment options to reduce suspended solids levels. The effluent quality will be much improved, and the frequency and cost of tankering sludge off-site dramatically reduced in many cases. Limiting factors are ground conditions, the nature of the effluent, sludge and related odours.

Clearly, a high water table or the possibility of polluting streams close to the surface will rule out lagoons unless effectively lined by butyl sheet or clay. This requirement may make construction uneconomic or impractical if the lagoon is ultimately to be bulldozed, or the sludge recovered by mechanical excavation.

As most lagoons are usually shallow and of considerable surface area, high annual rainfall which is only partially offset by evaporative losses may use up much of the available volume, and is a relevant factor in tropical areas. Lagoons have been successfully used to capture inorganic solids from the paper and cement industries, and in farming, where the odour potential of decomposing organic sludge is accepted, or remote from concentrated housing.

On the farm, in order to avoid gross hydraulic overloading during storms it is vital that the minimum amount of rainwater — particularly that collected from roofs — mixes with yard and animal pen runoff draining to a lagoon. A number of serious pollution incidents have occurred in recent years from lagoon banks bursting or over-topping, or the discharge being very high in solids and organic content because of minimal retention times.

If odour is not a concern, lagoons receiving organic sludges will act effectively as cold, slow digesters and may be ploughed up eventually. This practice is widely used in tropical rural areas as part of a complete sewage treatment system, where a number of lagoons are used in cascade and by rotation, the digested sludge being of considerable agricultural value and the effluent well purified and often low in pathogens. The high ambient temperatures are a benefit here, providing the sludge is kept covered by liquid.

Most lagoons are designed with retention times measured in days or weeks rather than hours to effect maximum solids removals, and a simple inlet distribution system at several points will help in spreading the sludge load evenly. The long retention time offers significant buffering capacity to attenuate periods of poor effluent quality, and the shallowness will assist mixing, neutralising reactions and coagulation in the industrial situation.

Ultimately, if the sludge is to be pumped out, 10% DM is a safe maximum for most materials, otherwise some form of dragline or excavator will be necessary. Cleaning out should be necessary only once every 1–2 years, and it could be carried out at much longer intervals than this. Hire of or investment in mechanical plant will be necessary, with associated transport costs. The recovery of the sludge for reuse or sale may be quite practical with careful working. Typical lagoon performance would be a sludge volume reduction by compaction or digestion of over 35%, and 40%—75% solids removal.

In sewage treatment, practice has been to use lagoons as a tertiary or polishing stage because they are simple, reasonably low in cost to construct and repair, and consume no power. In this application, their shallow construction assists ultraviolet (UV) penetration, and bacterial kill rates. If bacteriological standards are to be applied to sewage works effluents in the future, this will be a much cheaper option, given the space, than any method of artificial disinfection.

Rectangular Settlement Tanks

Uncommon now on sewage treatment plant except for holding storm flows prior to return, these tanks are easier to construct than the circular type and are quite common in a number of industries where the need to desludge the tank regularly is less important. A typical late-Victorian sewage works would have been blessed with these structures, constructed with elaborate brickwork and requiring manual desludging, neither a pleasant or particularly safe work activity.

Modern installations should be built with a ratio of width to length in the range 1:2 to 1:4, baffled inlet and a floor sloping slightly towards the effluent discharge end, to encourage sludge to gently roll into a well or discharge trough. A V-shaped floor (as seen in cross-section) with a central channel is an alternative arrangement.

Mechanical scrapers suitable for these tanks are more complex, requiring traversing and lifting functions. Most installations have no scraper set and rely on gravity to collect and withdraw sludge. If the tank is less than 2m deep, any baffles or scum boards fitted should only penetrate the top water level by 10-15cm to avoid turbulence and disturbing the underlying sludge layer by scouring. The baffle performs the important task of preventing influent liquid streaming across the top surface carrying floating debris to the effluent channel, and effects some mixing to reduce thermal stratification. In many situations, the incoming liquid is warmer than the tank contents.

Circular Settlement Tanks

This has become the standard design on sewage treatment plant, where construction in concrete has become a well practised art. Circular in plan, with a ratio of diameter to depth usually within the range 3:1-8:1, the tank base slopes from the periphery inwards by as much as 30°, encouraging the sludge to a central collection hopper, from where it can be withdrawn by pump or hydrostatic head.

Flat-floored circular settlement tanks are cheaper to construct, but uncommon, needing effective scraping. Most tanks are fitted with a mechanical scraping set. The effluent enters via a central pipe into a stilling box, flowing first down and outwards and then upwards to discharge over a peripheral weir. The tank is normally buried in the ground to top water level, requiring considerable initial excavation; this policy is dictated on many sites by the need to maintain a fall through the whole works so that all the flow proceeds by gravity, and costly intermediate pumping is avoided.

The long-term pumping costs of any elevated tank should always be considered in the budget for plant operation, as a poorly-laid-out site, where the disposition of the main components, including balancing and settlement tanks, is illogical or does not use a slope to advantage, will incur a constant energy cost. The construction quality, particularly of buried concrete tanks, is very enduring. They will be fit for decades of service, outliving the mechanical scraping equipment and other metalwork several times over.

High groundwater levels must be allowed for in buried tanks by providing flap valves in the tank to allow hydrostatic pressure relief when the tank is drained down. More than one sewage works in the world can boast a tank that has risen out of the ground by a measurable amount, and often wrecked pipework and electrical connections on route! Prefabricated settlement tanks made from grp or mild steel are more appropriate in many industrial areas, and can be installed at ground level very quickly.

Among suppliers of these, Environmental Engineering Ltd. is typical, offering an installation where a wedge wire clarifier can be fitted to the top and a sludge blanket generated. This additional feature acts as a filter, and 95% success rates in solids removals are claimed. Normal desludging is effected through a large gate valve at the base of the tank as required.

Two important considerations for above-ground tanks within the factory area are point loadings and sizes. A relatively small tank, 6m in diameter and 4m high excluding the hopper base, will weigh 115t when full. It is equally important to have adequate settlement capacity, pump sizes and inlet baffling to avoid disturbance. In many situations, a daily fill, settle and desludge system of working can be used. It may be possible to dovetail shift-working and effluent production so that the tank serves two different production areas and handles different effluents. Solids can then be kept separate and may be recovered and recycled.

Different effluents that may react and beneficially coagulate are best treated in an initial balancing or mixing facility upstream of the settlement tank. Circular tanks settling organic sludges requiring regular or constant removal are usually fitted with some form of scraper mechanism. There are a number of options and design varies with the type of sludge.

Fixed bridge scrapers have a stationary bridge spanning the tank, with the scraper supported and driven from the centre by a motor of 0.5-2.0 hp. Final drive is through a double reduction worm gearbox, and a special spur gearbox with a substantial thrust bearing supporting the suspended parts. The central drive shaft is located on the tank base in a spigot. About 20m in diameter is the largest practical size.

Half-bridge scraper sets are the more common, as they can be installed satisfactorily on tanks up to 45m in diameter, and there has been a trend towards larger settlement tank sizes. One end of the bridge is supported and pivots at the tank centre, while the other end is mounted on an end carriage that runs round the tank outer wall on a wheel (or rail in earlier designs). Contemporary installations have diminutive drive motors of about 300 W driving 1000:1 ratio gearboxes directly or via a chain.

As the bridge rotates, some allowance must be made for minor variation in the tank structure and wheel wear, so central pivot bearings are designed to take a small amount of vertical movement.

Power collection is via a slip ring collector at the tank centre, fed from either an overhead cable supported by catenary wire or by cable dueled up through the tank floor and stilling box. A shear pin is an essential component of the drive mechanism of either scraper type.

The type and design of scraper blade and scraper rotational speed are dictated by sludge type. Primary raw and heavy industrial semi-organic sludges are usually moved by blades hinged at the tank base and supported by rigid arms. A speed of 2.5 m/minute at the blade tip is possible without disturbance. Secondary humus and activated sludge are normally scraped with a blade hinged at the upper part of the arm; another hinge at the blade end can be provided and is probably the ideal trailing scraper, reducing blade wear and trailing weight, and giving considerable compensation for minor irregularities in the screed of the tank floor. A rotation speed of about 1.2m/ minute is a safe maximum for lighter, flocculent sludges. In both cases, the scraper blades are arranged in the form of a continuous volute, so that sludge is removed to the central well with every revolution around the tank.

Activated sludge tanks with a 30° floor slope often desludge satisfactorily with a simple chain scraper to disturb the sludge and encourage it to roll down to the central well. Tanks built with flat floors and settling light, secondary or flocculant sludges prone to denitrification are often fitted with V-shaped scrapers and individual sludge draw-off tubes. This arrangement, the sludge being removed by hydrostatic pressure. An air lift pump can also be used, the compressed air serving to scour the tubes clean and make them less prone to blockage.

A continuous scum baffle around the tank periphery, and about 30cm in- board of the weir, is an essential feature for all tanks where there is the slightest chance of floating material. This has become particularly important on secondary tanks, where penetration of plastics through conventional screens often results in their appearance at many sewage works after extensive secondary treatment.

On primary sedimentation tanks at sewage works, the scum baffle will prevent most floating grease particles and screenings from fouling the weirs and blocking filter arms, unless radical changes in upward flow rate sweep the material under the board. A board depth of 15cm into the liquid is recommended.

A scum collection box, slightly raised above top water level, can be provided, a small blade sweeping material collected against the scum board into the box with each bridge rotation. A hinged flap on the trailing side of the box depresses briefly as the blade passes, admitting tank water and flushing the scum into a discharge pipe.

To counter the floating plastic screenings problem on secondary tanks submersible pumps have been used to skim the top water surface and collect the material for discharge through a filter sack suspended over the central stilling box; this has proved successful. This operation on a settlement tank at an activated sludge plant. The volume of material in the sack is indicative of 24 hours collection and also reveals the magnitude of this problem in sewage treatment for materials that are now prohibited in effluent discharges to watercourses.

Desludging of mechanically scraped tanks settling primary, raw sludge is usually carried out every 5-12 hours, exceptions being dependent on the 'freshness' of the influent. A 4%-6% DM sludge is obtained. Periodic desludging allows some compaction and saves on storage and transportation costs.

The tendency for active secondary sludges, high in nitrogen, to float rather than settle has been mentioned, and a constant or very regular desludging regime is essential for humus or secondary settlement tanks, with sufficient storage to accommodate quantities of sludge rarely more than 4% DM. This is particularly relevant in Mediterranean and tropical climates where raw sludges quickly turn septic (if the sewage arriving at the works is not already septic) and high temperatures cause rapid denitrification of secondary sludges.

UPWARD FLOW CLARIFIERS

At its simplest, a wedge wire panel can be inserted into a settlement tank about 250mm below top water level and (by the selection of a suitable aperture size) it will act as a fine filter. This is quite often used as an alternative to building more tanks in overload situations; settlement performance can be improved due to two effects :

(1) UFRs vary quite widely through tanks, particularly those of the freestanding package variety; they are highest near the weir. Inserting a panel will equalise the hydraulic load across the tank and give better settlement generally, and especially of the larger irregular particles where the flow rate varies widely.

(2) Small particles or 'fines' whose settlement velocity is close to, or even less than, the upward flow rate will be trapped and tend to coagulate within the wire mesh, forming a blanket which will itself act as a filter. This effect can result in spectacular solids removals of over 95% for industrial effluents containing fibrous material.

The conventional clarifier, widely used for the primary treatment of surface waters during potable water production, generates a sludge blanket midway up a conical bottomed tank. Effluent is admitted through one or preferably several baffled inlets towards the tank base, and passes up through the sludge blanket, where a combination of physical trapping and coagulation ensures the high removal of solids customary for this system.

Aluminium sulphate, polyelectrolytes and granular activated carbon are all potential additives to assist flocculation, and doses can be determined experimentally. Fibrous solids will form a stable blanket without chemical aids, and the clarifier is widely used in the paper industry, where fibre recovery is economically essential.

The sludge is collected by cones suspended under water at the desired top sludge position. A maintained blanket thickness of 0.3-0.9m is typical with a rise rate of 3-5m/h. Automation is completed by optical sludge blanket detectors, which control sludge withdrawal rates and maintain the blanket position by adjusting the effluent weir height or incoming flow rate. A crucial aspect of clarifier operation is that large flow variations do not occur suddenly through the tank; upstream flow balancing or careful distribution where there are several tanks is therefore essential. As a guide, there is a 3:1 ratio between minimum flow that will just maintain the blanket, and the maximum which causes it to rise and expand suddenly.

The UFR for clarifiers is often quoted in the same range as that for settlement tanks: 0.5-1.0$m^3/m^2/h$. There are plenty of practical examples of clarifiers working periodically outside this range. The crucial point for those working at high rates is for the blanket detection system and sludge withdrawal facilities to be smart and quick enough to respond to the flow rate changes and so prevent a blanket loss.

In dimension, most clarifiers are built to a diameter/ depth ratio between 0.5:1 and 0.8:1, making a fairly deep or tail structure compared to footprint size.

In summary, clarifiers provide effective solids removal, but need quick and responsive control and stable flows to maximise potential.

Sludge Removal

Sludge contains very little dry material—some plants cannot do better than 0.5 per cent solid matter in the blowdown and 1-2 per cent can be considered quite good. Sludge disposal thus represents a sensible loss of water and chemicals, and it often creates an effluent problem. Sludge concentration is therefore important. (Sludge is not toxic, but it is an ugly discolourant. If it settles to the bottom of a lake or stream it can smother the natural organisms there, which must be avoided. The alternative to putting sludge into the drain is to lagoon it for further thickening and dumping. Obviously this becomes less expensive if the plant itself yields a sludge high in dry solids.)

A tank creates sludge continuously but the blanket stores it, so that blowdown can be continuous or intermittent. If blowdown is inadequate, sludge accumulates and eventually passes over the top and clogs the filters. If blowdown is excessive the only drawback is that the sludge concentration falls (or, in the limit, the whole blanket is destroyed). Control therefore boils down to blowing down at the lowest practical rate, and uneasy compromise between operator's convenience and the economist's and ecologist's requirements. One simple method is to have a sludge valve intermittently opened by a variable timer, set by the operator to match the plant's performance. Automatic control of de-sludging, for example by photoelectric cells, has not caught on because of the difficult conditions, in which most devices so far tried have failed. One interesting development (still in the development stage) is to collect the sludge in submerged containers which are emptied when their weight reaches a preset value. If this or any other automatic device is commercially successful it will result in higher sludge concentrations becoming practical.

A heavy sludge can thicken to a stiffish paste and cease to flow freely. Then a sedimentation tank will need scrapers to collect the sludge rather than relying on a sloping bottom to bring the sludge to the blow-off point. The blow-off lines themselves may need automatic flushing to keep them clear. Alum and iron flocs, however, do not usually form very stiff pastes.

SOME ASPECTS OF HYDRAULIC DESIGN

Inlet Systems

While inlet systems vary, depending on the type and shape of the sedimentation tank, the common design goal for all types of inlet system is to achieve a uniform distribution of flow over the flow cross-section. In plain sedimentation processes there is the additional requirement of dissipation of the kinetic energy of the incoming fluid so as not to disrupt the sedimentation process, while in fluidized bed processes there is the rather different additional requirement to inhibit settling of particles on the tank floor.

In rectangular primary tanks, the incoming flow is distributed over the tank width by a transverse manifold-type feed channel with multiple uniformly spaced discharge slots or orifices. The kinetic energy from these individual discharges may be locally dissipated by an appropriate baffle system.

In upward flow tanks and circular tanks, central stilling chambers are used to contain the turbulent mixing generated by the inflow which, as illustrated in these diagrams, is directed upwards towards the free surface. The stilling chamber diameter is typically (Christie and Harbinson, 1978; Metcalf and Eddy, 1991) within the range 12-20% of the tank diameter and extends from above the water surface to about mid-depth. If the stilling chamber extends too low into the tank, then the inflow may disrupt the settled solids in the central sludge hopper.

In sludge-blanket clarifiers, the inflow is directed downward towards the tank floor, thus preventing deposition of solids on the floor. In rectangular flat-bottomed tanks, pipe manifold distribution systems are used, with downward-discharging jets distributing the flow uniformly over the tank floor area. This uniformity of distribution, coupled with a corresponding uniformity of collection by closely spaced surface decanting channels, is essential to the maintenance of a stable sludge blanket in this type of clarifier system.

Outlet Systems

It is universal practice to take the outflow from the top of the tank. Weir outlets provide such a simple direct control of tank level that they are almost universally used. Single-sided launders or channels are be preferred to double-sided launders (Ekama and Marais 1986) since it has been found that the approach flow patterns associated with double-sided launders are more likely to cause erosion of the sludge blanket leading to solids loss. Kawamura and Lang (1986) found little difference in clarified effluent quality in a side-by-side comparison of equally loaded twin rectangular tanks, one of which was fitted with in- tank launders to give a weir loading of 10.4 m^3 h^{-1}, while the second tank had a simple transverse end weir resulting in a weir loading of 259.2 m^3 h^{-1}. Thus, there seems little empirical justification for the common design practice of providing in-tank decanting channels to limit the weir loading to a maximum value of about 10m^3 h^{-1} length of weir.

Plain weirs are unsatisfactory at low overflow rates because small variations in level result in uneven distribution of drawoff. For this reason it is general practice to use notched weir plates on sedimentation tank outlets. Circular tanks are typically fitted with a peripheral outlet weir and rectangular tanks with a transverse outlet weir. Flow in sedimentation tank decanting channels or launders is hydraulically designated as steady gradually varied flow; typically, such channels have a uniform lateral inflow over the channel length and have a free overfall at the outlet end, as illustrated in Figure 4.16. The water surface slope (dy/dx) in such a channel is described by the equation:

$$\frac{dy}{dx} = \frac{S_0 - S_f - \frac{Q}{gA^2}\frac{DQ}{dX}}{1 - \frac{WQ^2}{gA^3}}$$

where S_o is the channel bottom slope, S_f is the friction slope $=fv^2/8gR_h$, Q is the flow, $dQ/dX = q_L$ is the lateral inflow rate per unit length, W is the channel width, and A is the flow cross-sectional area.

Upper equation can be solved numerically, starting from a defined depth at the outlet end. The computer program DECANT calculates the variation in depth, y, along the channel, for channels of circular, rectangular and trapezoidal sections, assuming the depth at the outlet end to be the critical depth y_c, which is defined by

$$\frac{WQ^2}{gA^3} = 1$$

FLOTATION

We have described plant for sedimentation, which is the commonest process for removing floc between coagulation and filtration—until recently it was the only one available. Flotation has now become a possible choice for conventional water treatment applications and it can be considered a potential alternative to sedimentation. We still have to see, however, to what extent and in exactly which fields flotation will establish itself. The process has come into conventional water treatment via a number of applications in effluent treatment, and the picture is not made clearer by the fact that there are a number of different flotation processes for which differing and conflicting claims are made.

The Basis of Flotation

If we attach a gas bubble to a particle of suspended matter we can create a composite particle which is lighter than water arid will therefore float to the surface. The exploitation of this phenomenon started in the mining industry decades ago. It turns out that bubbles will attach more easily to some minerals than others, so that flotation is actually used as a selective process to concentrate finely ground ores. In effluent and water treatment, by contrast, it is essential that all the suspended particles should become attached to gas bubbles so that they all float out of the bulk of the liquid.

This means that the process plant must provide enough bubbles for at least one bubble to become attached to each particle. In turn this means that the number of bubbles provided, multiplied by the probability of a bubble becoming attached to a particle, must be at least equal to the number of particles in the water. There are two more potential variables which we can ignore in conventional water treatment, where we are dealing with iron or alum floe. One is the minimum amount of air necessary to float a single particle; these floes are so light that in practice a single bubble of the smallest size which can be readily produced will give enough buoyancy.

The other is the degree to which bubble and particle tend to stay together once they have been brought into contact, and again in the case of these floes this presents no problem. The important variables are therefore the number of bubbles and the probability of collision.

The cost of introducing bubbles into water is one of the main cost variables in flotation, and for a given process it depends largely on the mass of gas introduced. But the mass as such is not at all important; in this application we are interested in the number of bubbles. It is therefore advantageous to produce the smallest possible bubbles in order to get the largest possible number of them for a given mass of gas.

Small bubbles have the additional advantage that their terminal velocity (controlled by Stokes' Law) is small. This increases the time during which they are available for collision with floe particles, and when contact occurs the approach velocity will be moderate so that the chances of attachment are good. By contrast, very large bubbles are actually harmful: they rise so fast that the violent turbulence which they create can break up floe or knock bubbles off it.

Every argument therefore points to the plant being designed to create the smallest possible bubbles. (In theory it might be possible to create bubbles which would be too small but in practice this limitation can be ignored.) Moreover plant designers try to arrange for any large bubble which might form accidentally to be diverted out of harm's way.

Rise Rate in Flotation

The movement of the combined particle formed by a bubble attached to the floc particle is also controlled by Stokes' Law.

Once again the effective weight of the particle is $(d_1 - d_2)$ where fis the effective density of the particle and d_2 represents the density of water. Now that d_1 is less than 1, the term $(d_1 - d_2)$ becomes negative, to show the net force is upwards. The benefit of flotation results from the fact that the difference $(d_1 - d_2)$ is quite substantial and that particles therefore float to the surface much faster than floe sinks in sedimentation, where $(d_1 - d_2)$ has a very small (positive) numerical value.

Flotation is therefore a much more rapid process than sedimentation, with the result that flotation tanks are much smaller than sedimentation tanks of similar throughput. The total hold-up time in a flotation tank is typically about half an hour, and is a tenth to a quarter the time required in a conventional sedimentation tank.

Coagulation before Flotation

This short residence time has important consequences. Above all it means that there cannot be enough time for any significant coagulation or floe growth to take place in the flotation tank itself, whereas in conventional sedimentation tanks designers have plenty of time available, which they utilize in such devices as sludge blankets or sludge recirculation.

Flotation will only succeed, therefore, if the floe is fully formed by the time it enters the flotation tank. This means that a flotation plant needs a separate, carefully designed coagulation section preceding the flotation tank itself. In sedimentation, on the other hand, it is very common to let the tank perform both coagulation and sedimentation in one piece of equipment.

Coagulation Plant

We earlier described coagulation and its two main requirements. These are, first, that the chemicals and water must be mixed in such a way that the product (power input x residence time) reaches a certain minimum which is empirical. Second, this mixing must be carried out in such a way that the shear imposed on the water must not break up the floe as it coagulates, which means that it must match the floe strength at various stages of its formation.

Conventional purpose-built coagulation plant therefore consists of a flash tank in which water and chemicals are mixed quickly in high turbulence (which is possible because at the very first stage no floe has formed as yet), after which the water passes through a series of tanks fitted with large slow-moving paddle stirrers, with the stirring rate tapering off progressively as the floe becomes bigger. Ideally each of these tanks has an independently controlled variable-speed stirrer in order to give the greatest degree of control.

Depending on how difficult the water is to coagulate, there will be two, three or four of these stirring stages and the total hold-up of water in the coagulation stage may be up to half an hour. This means that the hold-up can be as great as it is in a flotation tank itself but the overall hold-up of the whole plant is still an hour or less.

The low hold-up compared with conventional sedimentation plant gives rise to the main differences between the two processes. It means that flotation plants respond much more rapidly to changes in conditions whether in the water or in the chemical dosage — this may or may not be an advantage, depending on the application. It also means that flotation can be used more successfully on an intermittent basis — a sedimentation tank will take hours after start-up before its treated water quality is satisfactory, while a flotation plant should be able to produce good water almost at once and will settle down to stable operation in about half an hour.

Experimental work shows that the empirical minimum value for (power input x residence time) holds quite well for alternative methods of mixing. This means that if satisfactory methods can be found of getting more power into water in a shorter time, then the hold-up of the coagulation section could be reduced. One device which works well on the pilot scale, for example, is a bundle of tubes and U-bends, in which the turbulence is created by reversing the direction of flow. It seems that this is one way of getting better mixing without excessive turbulence and so reducing the necessary hold-up. It also benefits from the fact that the water moves along a well-defined path, while in the mixing tanks there is always a lot of back-mixing. Unfortunately this idea is not particularly cost-effective and of course it provides no means of fine control, but it does show the lines on which we may see future progress.

Coagulation Chemistry

The chief object in coagulation chemistry, as far as flotation is concerned, is to get the smallest number of floe particles to be floated off, so that we need the smallest number of bubbles and in turn economize on the cost of introducing the gas into the water. As far as achieving the object of treating the water is concerned, of course, we also need a certain minimum coagulant dose in order to trap and remove the suspended matter and the organics.

It now becomes clear why sedimentation is reasonably tolerant of coagulant overdose but flotation is not. More coagulant produces more floe. In sedimentation the extra floe particles settle harmlessly enough; the penalties come only from using a little more coagulant chemical and creating a little more sludge. In flotation, however, the extra floe needs extra bubbles to float it off, which costs money.

Coagulation aids can be useful in flotation but not as useful as in sedimentation. They do a good job in promoting coagulation and reducing the number of particles to float off but their property of increasing the floe density is at best useless and can be positively harmful.

Flotation Processes

So far we have discussed flotation in general but there are many ways in which bubbles could be introduced into water, to result in processes with different characteristics:

- compressed air can be dispersed through fine orifices or sinters;
- a stream of water supersaturated with air at high pressure can be released into the bulk of water being treated to release the air as the pressure is taken off;
- electrolysis of the water can be used to raise oxygen and hydrogen bubbles on the electrodes; and
- the whole flow of water can be pressurized, supersaturated with air and the pressure then released.

Dispersed Air

This is the simplest, cheapest and least effective method. It cannot produce small bubbles, and it has no application in conventional water treatment.

Dissolved Air

At present this seems to be the most promising process for conventional water treatment. It is therefore the process we generally have in mind in the context of this book.

The normal design uses a side stream of clean water, most commonly recirculated clarified water, at a flow which is about 5-10 per cent of the throughput. (At start-up it may be best to use town mains or some external source of clean water.) This side stream is pumped at high pressure through some device in which air is dissolved in the water. Ideally, this device should bring the water near to saturation. If it does not, then it will be necessary either to work at higher pressures (involving higher cost) or to recirculate a larger flow in order to obtain the necessary mass of gas to yield the necessary number of bubbles.

A conventional packed tower for dissolving air is capable of achieving near-saturation. It is supplied with compressed air from a receiver and compressor, which maintains the necessary pressure and automatically makes up the amount of gas taken away by the water. There are other devices which use some kind of direct injection of air, but on the whole these do not achieve the same degree of saturation, though they are cheaper to build.

Float tanks The float tanks using dissolved air do not vary in principle. They are, on the whole, rather simple tanks of about 1.5m depth, with a water path of perhaps 4m. The width of the tank is made to vary according to the throughput rating; alternatively some tanks are circular with a centre inlet and a peripheral outlet.

One ingenious design places the float tank above a conventional deep filter bed so that the clarified water sinks straight through the bed. This means that the flotation tank effectively becomes the filter rising space, and, when the filter is backwashed, the whole flotation plant is washed with it.

The cross-section of a typical flotation tank. The water inlet and outlet are both near the bottom of the tank; the inlet has to be engineered carefully to give good distribution across the tank without creating turbulence which could break up the floe. Usually there is a sloping baffle near the water inlet to guide the water towards the surface. The recycled and saturated water is fed into the main water stream through a series of nozzles. The type and number of nozzles vary considerably between different designers' plants. The nozzle must be capable of producing very fine bubbles without transmitting undue turbulence to the bulk of the water, and it must withstand clogging and corrosion/ erosion. Some designs use fixed orifices to release the high-pressure recycle; others use a modification of the needle valve.

The positioning of the nozzles must ensure good distribution of the bubbles through the bulk of the water and they must go near the bottom of the tank in order to maximize the possible time for contact with floe particles as the bubbles rise to the surface. If there is any tendency for large bubbles to form, these should be taken away at a place where they cannot interfere with the float which accumulates on the surface of the tank. The tank must not have any overhanging obstructions under water level which could accumulate air-laden floe. In time, such accumulations release their air and fall back into the tank in large lumps which cannot be re-floated and which will therefore pass on with the clarified water.

For good distribution, there is usually a nozzle about every 30 cm of tank width. One convenient way of controlling the flow of recycle is to treat the nozzles as fixed orifices even if they are in the form of needle valves and to vary the water and air pressures supplying the flow.

Electroflotation

Here the bubbles are created on electrodes in the bulk of the water; the plant therefore needs pairs of electrodes with a small, fixed gap between them, through which the DC passes, using the water as electrolytic medium. The ohmic resistance of the water represents a major factor in the power cost and the process therefore becomes very expensive in electrical power unless the water has a fairly high conductivity. This characteristic makes electroflotation more suitable for effluent than for conventional water treatment.□The DC is produced by a transformer rectifier w ose cost is roughly proportional to the current required. The electrodes also follow a similar cost pattern. This pattern is different from the bubble-producing gear in dissolved air plant, where the cost of the dissolving plant might vary only as the square root of the mass of air to be dissolved, so that for the very small flows the cost becomes proportionately large. Electroflotation might therefore find application in very small plants for which dissolved air gear becomes disproportionately expensive.

Electrodes are capable of producing very small bubbles in unlimited quantity and without causing any turbulence. This is no particular advantage in conventional water treatment where only a moderate amount of floe has to be removed but it makes electroflotation a good prospect in effluent problems where a very large number of bubbles is needed.

Different designs use all kinds of electrodes. Electrodes are subject to severe corrosion and must either be made as consumable spares, or use expensive materials of construction. Either way the electrodes are a serious cost factor.

Flotation tanks for this process can resemble tanks for dissolved air processes, but there is a large variety of designs.

The Static Head Process

This process uses a shaft about 10m deep, notionally arranged as a U-tube (see Figure 4.18). The water flows in and down one leg and up and out of the other; at a depth of 10m the water is then under about 1 bar static head, but the pumping cost is minimal.

An excess of compressed air is dispersed into the water near the bottom of the down leg; the bubbles rise against the downcoming water and dissolve under the static head. Any surplus air rises to the surface and is lost.

The supersaturated water tends to lose dissolved air as the static head falls as water rises up the upflow leg. Bubbles form on the nuclei provided by suspended particles in the water, and therefore this process effectively avoids the problem of generating fine bubbles and bringing them into contact with the suspended matter. The principle is rather elegant, but practical limitations seem to have restricted this process to treating biological wastes.

Sludge and Sludge Removal

The sludge forms as a float on the surface of the flotation tank. Compared with sedimentation, the solids content of the float is high, partly because of the greater force pushing the rising particles into the accumulated mass of sludge and partly because air bubbles in part replace the interstitial water.

In the float there is a tendency for the small air bubbles to coalesce in time and leave portions of sludge without attached bubbles. This does not matter as long as the float is not disturbed because the fresh-floating particles continue to come up from underneath and hold up the whole layer of float. While the sludge is being removed, however, there is also a tendency for bits of sludge to detach themselves, sink back into the tank and pass out with the clarified water. Sludge removal must therefore be done with care.

Most designs have continuous scraper gear skimming sludge off the surface and on to an inclined beach, from which it drops into a hopper or trough. The skimming rate is adjusted so that the float builds up to a thickness of 2-3 cm; on the one hand this permits the sludge to be taken away at a reasonable solid concentration (at least 1 per cent) but on the other it means the float has not been left long enough for a serious amount of material to drop out when it is disturbed. Another method is to allow the sludge to build up and periodically to raise the water level in the tank so that it overflows a weir, carrying the sludge with it.

In some conventional water treatment operations the coagulant dose is quite small (especially when compared with the amounts used in effluent applications). The sludge then accumulates quite slowly. Where this is the case, it is possible to let the float accumulate for several hours at a time and remove the sludge on an intermittent basis. If the float can be left to accumulate for more than eight hours, say, it might even be possible to control desludging manually and save the expense of automatic gear. The float becomes extremely stiff and may be up to 10 per cent dry so that it will not flow easily and may need helping along, especially near the walls of the tank (to which the float tends to adhere). This can be done with water sprays, or manually. This kind of desludging leads to a slight loss of quality in the clarified water for ten or twenty minutes, but at such infrequent intervals that the additional load on the following filters is effectively very small.

Flotation vs Sedimentation

We might now compare the dissolved air process (which is the main contender for conventional water treatment applications) with the conventional sedimentation tank which is at present the standard apparatus for floe removal.

The main difference between the two processes is the short hold-up time in flotation. Flotation therefore uses much smaller tanks and its capital cost overall ought on that account to be lower. It takes up much less space, and for small sizes can be built skid-mounted or even transportable. On the other hand, the float must not be exposed to rain or wind, both of which might break it up, so that the plant must be housed in a light shelter at least.

The short hold-up makes flotation plants much more responsive to changes in conditions, as already mentioned. On a water liable to rapid changes, such as a small Scottish stream, this could mean almost continuous manual attention to the plant when a period of rain follows one of drought. A sedimentation plant in the same circumstances is also likely to give trouble but it will not require such continuous attention for the simple reason that its response time to changes is so slow that there would be very little point in sitting on it continuously.

On the other hand, flotation plants can be started and stopped almost without delay, which might be very important in suitable circumstances. While flotation plants may show a lower capital cost for performing a given task, they inevitably show a higher operating cost. This is due to the cost of high pressure pumping of the recycle and air. There may be some small saving due to the fact that flotation plants may be thriftier in coagulation chemicals than sedimentation plant, but this will not cancel the increased cost due to pumping.

Rotation can produce as good a quality of clarified water as sedimentation, or sometimes even better. 'Can' means that the comparison is made between the products of two well-operated plants. It does not take into account the fact that in many cases, sedimentation plants can tolerate a greater degree of the ignorance, neglect and maloperation which is unfortunately the fate of many water-treatment plants.

When well operated, flotation is capable of giving a better product when treating the difficult thin, cold, coloured waters of Scotland and such countries. This is why the use of flotation for water treatment received much of its impetus in Scandinavia where these waters are common. On the other hand, waters of high turbidity are more suited to sedimentation. Algae removal (which is difficult by any other means) is performed rather satisfactorily by flotation.

Flotation produces a smaller bulk of sludge, and with care is capable of producing a very dry sludge. In some cases this can be a very important benefit.

Summing up: flotation is a relatively new process, especially in the field of conventional alum and iron floe removal when treating natural waters. It is still not quite clear to what extent it will find an economical application in the field in which sedimentation has served at least adequately for so long. Algae removal and the treatment of thin, coloured waters by flotation both look promising. It seems as though the dissolved air process is the most suitable but, as the process is still being developed, something new may yet emerge.

3

WASTE WATER TREATMENT BY BIOLOGICAL PROCESSES

INTRODUCTION

Domestic sewage is a mixture of waste waters arising from household waste facilities including toilets, bathrooms and kitchens. It contains pathogenic microorganisms (bacteria and viruses that cause disease) and creates odour nuisance (sulphurous smell). Most sewage-treatment plants are designed to purify domestic sewage to an effluent quality adequate for discharge to the nearest watercourse. An exception exists in arid climates where waste waters are often used for landscape irrigation and other industrial and agricultural purposes.

Sewage is typically about 99.9 per cent water; the other 0.1 per cent is mainly organic materials (organic means derived from living things), a wide range of compounds comprising the elements carbon, hydrogen and oxygen plus smaller amounts of nitrogen and sulphur. The organic impurities enter the water during its various household uses.

The majority of the organic material is biodegradable, that is it can be consumed by microorganisms (bacteria etc.) and thereby removed from the water. The materials are converted to more microorganisms (which appear as sludge), carbon dioxide (which is a gas that escapes to atmosphere) and soluble salts including nitrate and sulphate. The high biodegradability of the organic material allows biological processes to be used to purify the sewage.

The biological treatment of water involves various types of fermentation. Fermentation is the deterioration of certain organic substances and is often accompanied by the release of gases due to the action of the enzymes chat are secreted by microorganisms.

GROWTH OF A BACTERIAL CULTURE

After being seeded, a bacterial culture continues to grow until the nutrients in the medium are used up, provided that the environmental conditions are favourable. Figure 10.1 illustrates the variation of X, the concentration of a bacterial culture, in terms of mass per unit volume as a function of time t, under constant conditions of temperature, pH, etc. Several phases occur in succession:

- **Lag phase** : During this acclimatization phase, the cell synthesizes mainly those enzymes that are required for the metabolization of the substrate. This phase is especially important when the water has not previously been seeded with suitable micro-organisms. This may be the case in certain industrial effluents. During this phase, there is no cellular reproduction.

  $$X = C = X_o$$

  Where, X_o is the cellular concentration at time t = 0: the growth rate is therefore zero:

  $$\frac{dX}{dt} = 0$$

- **Exponential growth phase** : This phase occurs when the cellular reproduction rate reaches its maximum and remains constant in the presence of a non-limiting concentration of the substrate. This phase is measured by the generation time t_g (or doubling time), which corresponds to a doubling of the bacterial population, which thus has its minimum value. During this phase, the growth rate, dX/dt, increases in proportion to X, resulting in the exponential form of the curve. In semi-logarithmic coordinates, the curve takes the form of a straight line with the following equation:

$$\frac{dX}{dt}.\frac{1}{X} = \mu_m$$

where μ_m is the maximum growth rate. Or:

$$\text{Log}\frac{X_2}{X_1} = \mu_m\left(t_2 - t_1\right)$$

where t_g is given for $X_2 = 2X_1$:

$$t_g = \frac{\text{Log}_2}{\mu_m} = \frac{0.693}{\mu_m}.$$

- **Declining growth phase** : This phase comes about with the depletion of the culture's medium and the disappearance of one or more elements necessary to the bacterial growth. In some cases, growth is slowed by the accumulation of inhibiting products resulting from bacterial metabolism. X continues to increase, but dX/dt decreases.

- **Stationary phase** : X reaches its maximum value, X_{max}. Growth comes to a halt even if the cells maintain some metabolic activity.

- **Endogenous phase** : The concentration of living cells decreases because of an increasing mortality rate. The enzymatic autolysis of the cells causes them to die.

These various phases and the equations that govern them are applicable to aerobic and anaerobic media. The values of the various coefficients depend, of course, on the nature of the microorganisms, the substrate, and various factors, such as temperature and pH.

BACTERIAL GROWTH MODELS

In most industrial applications, the biomass is in the declining growth phase because the required levels of pollution control result in weak final concentrations of the substrate. Several mathematical models have been proposed to integrate the part of the curve corresponding to the declining growth phase and beyond. However, most perfected models attempt to cover the total growth curve.

Monod's model is the oldest, the best known, and the most widely used. It is an empirical model, very close to the Michaelis-Menten equation for enzymatic reactions. It is expressed as follows:

$$\frac{dX}{dt} = \mu X$$

where $\mu = \mu_m \cdot \left(\frac{S}{K + S}\right)$

where S is the concentration of the solution in the growth-limiting substrate.

K signifies a concentration threshold below which the growth rare becomes closely dependent on the concentration of substrate. This constant corresponds to the value of S for which $\mu = \frac{\mu_{max}}{2}$.

K is usually very small, and during the entire exponential growth phase, $\mu = \mu_{max}$.

It must be emphasized that in the case of bacterial growth involving other strains of microorganisms, the measured value of μ is in fact a resultant μ. At any given moment, some bacteria may be in a different growth phase from other bacteria.

Other models are used for specialized studies such as pure cultures and the use of a merabolite.

THE ACTIVITY OF A BIOMASS

In every fermentation process the mass of microorganisms present is an important factor, but another equally important factor is their activity. The goal is to optimize the product mass of microorganisms x specific activity.

Several methods have been suggested to measure the activity of a biomass. The ATP (adenosine triphosphate) measurement has been all but abandoned by water treatment specialists because the results are difficult to interpret. The measurement of the dehydrogenase activity is hindered by the dispersion of the results, while the measurement of DNA (deoxyribonucleic acid) involves a long and painstaking dosing process.

Respirometric methods are used most often.

ELIMINATION OF THE SUBSTRATE

With a given mass of microorganisms, the elimination of substrate S, as a function of time, may occur at different rates.

The concept of reaction order is generally used. A reaction is said to be of zero order if dS/dt is constant, which means that the reaction rare is independent of the substrate concentration. The opposite of this is a relation of the type $dS/dt = K.S^n$, where the reaction is said to be of nth order.

In water treatment, the development of S as a function of time, for a given constant mass of microorganisms. Initially, the reaction is of zero order. Once the substrate concentration drops below a certain value, the order of the reaction changes, *i.e.*, the rate at which the substrate disappears drops. This means that the final substrate fractions are often difficult to eliminate.

The first phase corresponds to complex biosorption and flocculation of colloidal matter phenomena. Subsequently, the disappearance of substrate is linked to its interaction with the microorganisms.

AEROBIC AND ANAEROBIC GROWTH

Biological processes used in water treatment applications simply harness natural phenomena. There are two ways of controlling removal of pollutants from effluents:

- with aerobic processes when oxygen is involved in the reactions. These processes occur spontaneously in water that is sufficiently aerated. Organic carbon is transformed into CO_2 and biomass;
- with anaerobic processes when the reaction cakes place without air, in a reducing medium. After degradation, organic carbon exists in the form of CO_2, CH_4 and biomass. Because of the low redox potential, nitrogen occurs in the shape of NH_3, and sulphur in the form of H_2S or the various types of organic sulphur compounds, such as mercaptans.

It is usually possible to illustrate the degradation of glucose in these two processes:

$$C_6H_{12}O_6 \xrightarrow{\text{in aerobic conditions}} 6CO_2 + 6H_2O + 2.72\ kJ.mol^{-1}\ (650\ cal.mol^{-1}).$$

$$C_6H_{12}O_6 \xrightarrow{\text{in aerobic conditions}} 3CO_2 + 3CH_4 + 0.144\ kJ.mol^{-1}\ (34.4\ cal.mol^{-1}).$$

In the synthesis of one gramme of biomass, which requires the same input of energy whichever process is used, the rate of generation of aerobic nuclei is higher than that of anaerobic nuclei, and the degradation process of the carbonaceous matter is more rapid. On the other hand, less sludge is produced in anaerobic conditions.

The term anoxia is generally applied only to an environment practically devoid of dissolved oxygen, bur in which acid and anaerobic fermentation does not occur.

TOXICITY AND INHIBITION

Successful fermentation, whether aerobic or anaerobic, requires rather stringent conditions concerning the medium. The temperature and the pH play particularly important roles. Equally important, the medium must not contain any toxic products or inhibitors that may slow the process or irreversibly hair bacterial activity.

Most heavy metals act as toxins on bacterial flora. This is especially true of copper, chromium, nickel, zinc, mercury and lead. These metals act by attaching themselves to certain enzymatic sites and blocking them, or by denaturing certain enzymes or, lastly, by changing the permeability of the cellular membrane.

Anions such as cyanides, fluorides, arsenates, chromates and dichromates all behave in a similar way. Halogens and certain organic compounds may also denature proteins or other cell components. Bacteria are not equally sensitive to different toxins. Their sensitivity to a given product also depends on their physiological state. Some strains are even capable of degrading toxins such as cyanide and phenols.

In practice, it is often possible to acclimatize a bacterial growth to the presence of toxins or inhibitors and thus lower the toxicity thresholds. Certain metals attach themselves to bacterial floc in the form of insoluble organometallic compounds without disturbing bacterial growth. It should be noted, however, that excessive levels of these metals in sludge may render the sludge unsuitable for agricultural use. Above a certain threshold, certain metabolites may themselves inhibit bacterial activity.

SALIENT FEATURES OF BIOREACTORS

Suspended Growth and Attached Growth

Bacterial growth may be used in many different ways. Traditionally, a distinction is drawn between "suspended growth" and "attached growth" processes.

- Suspended growth processes, used solely for treating wastewater, stimulate the growth of a bacterial culture dispersed in floc form in the liquid being treated. The culture is kept in suspension in a stirred tank in which one of the following conditions exists:
  - a given concentration of oxygen, in the case of aerobic processes such as activated sludge and aerated or natural lagoon processes;
  - the exclusion of oxygen, in the case of anaerobic processes such as contact-clarification, sludge blankets, and anaerobic lagoon processes.
- Attached growth processes make use of the ability of most microorganisms to produce exopolymers, which enable them to become fixed to widely varying supports so as to form a biofilm. Attached growth, like suspended growth, may be used in either aerobic or anaerobic treatment (fine granular media biofikers, trickling filters, biological discs, etc.).

Suspended growth has the basic advantage of being easy to use. However, since the concentration of microorganisms cannot exceed certain limits, suspended growth processes require structures that can hold large volumes of liquid.

Reactors can be smaller when attached growth is used, because higher concentrations of biomass, and at times higher activity levels, can be obtained.

It should be noted that in suspended growth as well as in attached growth there is an excess production of biomass, which must be extracted, treated and disposed of.

Dispersion and Hydraulic Retention Time

Another important distinction, especially in suspended growth processes such as those involving activated sludge, must be made between the homogeneous or completely mixed bioreactor and the heterogeneous bioreactor, such as the plug flow type. A bioreactor is said to operate as completely mixed when the concentrations (biomass, substrate, oxygen, etc.) and the temperatures are identical throughout the reactor. In plug flow bioreactors, the channels have a high length-to-width ratio and axial concentration gradients exist.

In an ideal plug flow reactor, all the particles entering the reactor at a given

moment are thereafter continuously subject to identical hydraulic conditions (velocity and direction). The hydraulic retention time is theoretically the same for all the particles, whereas in a completely mixed reactor the distribution of the retention times for the particles is Gaussian.

These hydraulic factors are important because they may affect the reaction kinetics and may even encourage the growth of particular bacterial species.

- **Reaction Kinetics** : Theoretical considerations indicate that, for a unit containing a given volume and when the reaction order is greater than zero, the reactions in a plug flow bioreactor will be more advanced than those in a completely mixed bioreactor, in other words, that plug flow reactors can be smaller than completely mixed units.
- **Dispersion coefficient** : On an industrial scale, there are no bioreactors that are strictly of either the completely mixed or the plug flow type. However, some units are approximately equivalent to one type or the other. In order to measure just where the units are positioned in this respect, a longitudinal dispersion coefficient is used. A value of zero signifies a perfect plug flow, while an infinite value corresponds to a perfect complete mixing.
- **Hydraulic retention time** Given that: t_t is the theoretical retention time given by $t_t = V/Q$, where

V is the volume of the reactor and Q is the flow, and t_r is the weighted average real retention rime, the response curve may be drawn after instantaneous injection of a tracer.

A number of cases may arise:

- $t_t < t_t$

  This case is an indication of the dead zones in which stagnant water plays little or no part in the various reactions. This phenomenon is usually reflected in the curve by a decline.

- $t_t = t_t$.

  In this case the complete geometrical volume is traversed by water.

- $t_t = t_t$.

  This result indicates that there is a short-circuit inside a reactor, if t_t is the average retention time of water having actually passed through the reactor.

  The use of bioreactors in the treatment of water that operate under increasingly intensive conditions requires that all these hydraulic concepts be taken into consideration.

PROPERTIES OF THE SUBSTRATE

A substrate is a group of products that is contained in water and is liable to be used by bacteria for growth. These elements may be classified in the following way:

- major elements: C, H, O and N;
- minor elements: P, K, S and Mg;
- vitamins and hormones;
- trace elements (Co, Fe, Ni, etc.).

In the particularly complex environment of most types of wastewater there are usually sufficient concentrations of the trace elements, vitamins and hormones for proper purification to cake place. The same is true for K, S and Mg. On the other hand, there may not be enough phosphorus or even nitrogen, in which case they must be added. These elements may have to be removed, in order to combat eutrophication.

In order to treat an effluent by biological means, it must have properties that are compatible with bacterial growth: suitable pH and temperature, no inhibitors or toxins.

Carbonaceous Pollution

Organic carbon is usually the main pollutant that must be removed. Organic carbon is also the principal constituent of the biomass, a highly simplified formula being $C_5H_7NO_2$. Because there are so many different forms of carbonaceous pollution, it is usually described in terms of global characteristics.

The changes in BOD versus oxidation time. About twenty-one days at 20°C are needed for complete oxidation, at which point the ultimate BOD, or the BOD_{21}, is obtained.

If all the organic matter in a water was biodegradable, then: $COD = BOD_{21}$

This is the case for glucose, where:

$$\frac{BOD_{21}}{BOD_5} = \frac{COD}{BOD_5} = 1.46$$

When there is non-biodegradable organic matter, as in the case of domestic wastewater and many types of industrial waste, then:

$$COD > BOD_{21}$$

Examples of non-biodegradable organic substances are cellulose, lignin, tannins, sawdust, etc.

During a biological treatment process, the COD/BOD_5 ratio of the effluent increases substantially.

Nitrogenous Pollution

Practically all sources of organic and inorganic nitrogen can be used by various microorganisms. When metabolized, nitrogen produces essentially proteins, nucleic acids and the polymers of cell walls. Nitrogen can be said to represent approximately 12% of the dry weight of a pure biomass. In the case of wastewater treatment, this value usually drops to below 10%.

The following types of nitrogen are found in wastewater:

- reduced forms, which correspond to Kjeldahl nitrogen: organic N, $N\text{-}NH_4^+$ (ammonia nitrogen).
- oxidized forms: $N\text{-}NO_2^-$ (nitrous nitrogen), $N\text{-}NO_3^-$ (nitric nitrogen).

The term "total nitrogen" is often used in wastewater treatment to refer to the sum of all reduced and oxidized forms.

Phosphorous Pollution

Phosphorus is present mainly in nucleic acids, phospholipids and polymers of bacterial walls. In certain cases, it may be stored in the cell in the form of polymetaphosphate (for biological phosphate removal. A very small fraction of the phosphorus is in the form of diffusible organic phosphorus, such as ATP (adenosine triphosphate).

Phosphorus accounts for 1.5 to 2% of the dry weight of a biomass. It should be noted, however, that this percentage rises with the growth rate and varies inversely with the temperature.

Phosphorus may be present in wastewater either in the form of orthophosphate, polyphosphate or organic phosphorus. Similarly, the term "total phosphorus" is used to refer to the sum of all forms of phosphorus.

POLLUTION AND THE RECEIVING MEDIUM: THE IMPORTANCE OF MODELS

The consequences of discharging pollution in a receiving water can be estimated by direct, on-site measurements, but this is usually performed for only a very limited range of hydraulic conditions.

On the other hand, in order to project the influence that future discharges or accidental spills would have, or to set up a programmed reduction of pollution discharges, the use of mathematical models is indispensable. The models currently used are based on:

- the re-oxygenation of the aqueous environment (from the air and by photosynthesis of aquatic plants and algae);
- the oxidation of organic matter and of reduced forms of nitrogen.

The purpose of these models is to determine the changes in the levels of dissolved oxygen and in BOD_5 (or TOC), $N\text{-}NH_4^-$ etc., downstream of the point of discharge.

The formulae proposed are almost entirely based on experience. The reaeration constants in river water depend mainly on the velocity and the depth of the watercourse. Using models of reduced forms of nitrogen remains risky because of the uncertainty in defining the polluted flows (especially those from agricultural operations) as well as the complexity of nitrification phenomena. Substantial difficulties arise when the changes in river sediments are taken into account.

Models have probably been most successful in the field of eutrophication of lakes.

BIOLOGICAL PERSPECTIVES

Progress in biotechnology is evidenced by the appearance of new techniques in enzymatic, immunological and genetic engineering. These new techniques hold great promise for the water treatment sector, even if some of them remain to be perfected or may appear unrealistic because of the complexity of the substrates to be treated.

One interesting method is to use the specific features of the enzyme either to accelerate the degradation of particular substances that are present in high levels or to develop biological sensors or probes. These systems, based on the attachment of enzymes or microorganisms to a membrane, should provide for rapid and selective measuring of pollutants or micropollutants (such as pesticides, etc.). Other techniques will be able to make use of methods of immunity recognition to detect microorganisms even in very low concentration (for the treatment of drinking water).

It is also tempting to use biotechnology to generate microorganisms capable of removing the pollution from effluents with kinetics that are much higher than normal levels, or to degrade substrates that have thus far eluded conventional methods.

However, the production and use of these "mutant" bacteria on an industrial scale also pose some difficult problems:

- competition with natural microorganisms;
- their behaviour in the presence of the large number of substances to be degraded;
- leakage of these microorganisms into the natural environment.

On the other hand, isolating and producing microorganisms taken from the natural environment so as to introduce large numbers into environments where they are lacking, may be a means of accelerating a selection and purification process that would otherwise be too slow.

AEROBIC BACTERIAL GROWTH

Aerobic treatment facilities have long been designed empirically by rule of thumb: 150 to 200 litres of aeration tank per population equivalent or 100 litres of trickling filter per population equivalent, etc.

A more rational approach to planning and designing such facilities is now possible thanks to:

- on the one hand, the benefits of microbiological studies, which, by explaining many phenomena, have perfected the original treatment processes for removing not only carbonaceous pollution, but also nitrogenous and phosphorous pollution;
- on the other hand, the development of fermentation technology that makes it possible to determine more precisely the principal parameters of a facility, such as hydraulic circuits, tank capacity, oxygen requirements, sludge production, etc.

Suspended Growth (Activated Sludge)

Definitions

Activated sludge processes essentially involve a phase in which the water to be purified is brought into contact with a bacterial floc in the presence of oxygen (aeration), followed by a phase of separation from this floc (clarification). In fact, these processes amount to an intensification of the phenomena that occur in the natural environment. The difference lies in the greater concentration of microorganisms which results in a greater oxygen volume demand. Moreover, in order to maintain the bacterial mass in suspension, it must be artificially mixed.

The birth of this process can probably be traced back to Friday, April 3, 1914, when two British researchers,

Edward Ardern and William Lockett, presented an account of their work, entitled "Experiments in the Oxidation of Sewage without Filters" to the Industrial Chemical Society of London.

Until the end of World War II, purification facilities were very modest in design. It was not until later that high rate systems using combined tanks (Aero-accelerator, Oxycontact, Oxyrapid) and the "Biosorption or Contact Stabilization" processes, etc., were developed.

Currently, research into more efficient purification processes prompted by the frequent need to remove nitrogen and the relative difficulties involved in operating high rate systems, have reawakened interest in low rate treatment.

- a so-called aeration tank, in which water to be purified comes into contact with the purifying bacterial mass;
- a clarifier, in which the purified water is separated from the bacterial growth;
- a recirculation device for the return of the biological sludge from the clarifier to the aeration tank. This arrangement enables the tank to support the quantity or concentration of microorganisms required to maintain the desired level of purification;
- a device for the extraction and disposal of excess sludge, or surplus bacterial growth, which is permanently synthesized from the substrate;
- a device supplying oxygen to the bacterial mass in the aeration rank;
- a stirring device in the aeration tank that guarantees optimal contact between bacterial cells and the nutrient, prevents deposits and promotes the distribution of oxygen to all the areas where it is needed. The same unit is very often used for both aeration and stirring.

The sludge suspension in the aeration tank containing the purifying bacterial flora is called **activated sludge**.

The way in which a treatment facility using activated sludge is fed is an important parameter. There are various methods.

Basic Relations For the Removal of Carbonaceous Pollution

These relations involve the use of typical coefficients that depend on the nature of the substrate and the physiological stare of the biomass (sludge age).

Oxygen Requirements and Excess Sludge Production

While the biodegradable organic matter is consumed by a mass of microorganisms under aerobic conditions, the following occur:

- on the one hand, the microorganisms consume oxygen to satisfy their energy demand, their reproduction by cellular division (synthesis of living matter) and their endogenous respiration (auto-oxidation of their cellular mass);
- on the other hand, a surplus of living matter and inert matter is generated, which is called **excess sludge.**

It is difficult co determine experimentally the active concentration X_a of activated sludge. However, it is possible to measure the concentration of volatile substances X_v and that of total SS (organic and inorganic), X_t.

To illustrate these various phenomena, glucose may be used as an example of the degradation of a totally biodegradable molecule. In the first stage, additional assimilable nitrogen transforms the glucose into cellular protein, whose formula can be represented as $C_5H_7NO_2$. In the second stage, this protein is degraded inside the cell itself to provide the energy required to sustain the cell. These two reactions can be expressed by:

Synthesis

$$6C_6H_{12}O_6 + 4NH_3 + 16O_2 \rightarrow 4C_5H_7NO_2 + 16CO_2 + 28H_2O$$

Auto-oxidation or endogenous respiration

$$4C_5H_7NO_2 + 20O_2 \rightarrow 20CO_2 + 4NH_3 + 8H_2O$$

These two reactions clearly both occur in a purification plant, but the latter never achieves completion because the requisite retention time of the sludge would demand extremely large volume tanks.

Even though the second reaction is never fully completed, it does occur to a varying degree, depending on the processes used. The higher the degree of completion, the less excess sludge is produced but the more oxygen is consumed.

In the above example, the complete oxidation of 6 molecules of glucose required 36 molecules of oxygen. These 36 molecules correspond co the COD of 6 molecules of glucose, or the ultimate BOD. Of the 36 molecules of oxygen, 16 were used for synthesis and 20 for endogenous respiration.

The fraction of the ultimate BOD used for synthesis, a'_u, is defined as:

$$a'_u = \frac{16O_2}{36O_2} = 0.45$$

The fraction of the ultimate BOD used for the complete oxidation of living matter, a_u, is defined as:

$$a_u = \frac{20O_2}{36O_2} = 0.55$$

The parameter a_{mu} may be compared to cellular efficiency and corresponds to the mass of cells formed by the mass of the ultimate BOD removed. In the preceding case:

$$a_{mu} = \frac{4C_5H_7NO_2}{36\,O_2} = 0.39$$

Thus, when degrading 1 g of the ultimate BOD, 0.39 g of living matter is synthesized.

Application to determine oxygen requirements

There are two sorts of oxygen requirements:

- oxygen required for bacterial synthesis, which is expressed by:

  a'_u x ultimate BOD removed

- oxygen required for endogenous respiration. As already mentioned, the entire mass of synthesized bacteria is not oxidized into CO_2 and H_2O. Only a fraction b_u of the synthesized $4C_5H_7NO_2$ is transformed into CO_2 and H_2O. In other words, only a fraction b'_u of the $20O_2$ required for the complete oxidation of the living matter must be furnished:

$$b'_u = \frac{b_u}{4C_5H_7NO_2} 20O_2$$

Oxygen requirements for endogenous respiration may be expressed by the formula:

b'_u.mass of living matter

Thus, the overall requirements are:

a'_u.ultimate BOD removed + b'_u.mass of living matter.

For the purpose of calculations, the oxygen requirements are expressed in kg per day.

For greater convenience, the following values are generally used:

- a' referring to BOD; and not the ultimate BOD;
- b' referring to the mass of volatile substances (and, at times, the total mass) and not to living matter.

Application to determine the production of excess biomass

The production of excess biological sludge is affected by two factors:

- the production of biomass during synthesis reactions;
- the consumption of a part of the biomass during endogenous respiration reactions.

The biomass produced during synthesis is expressed by: a_m.BOD; removed.

Endogenous respiration consumes: b.mass of VS. Thus the balance is: a_m.BOD; removed - b.mass of VS.

Sludge production is expressed in kg of SS per day. The amount of nonbiodegradable suspended solids contained in the raw influent must be added.

For greater convenience, the coefficient a_m is expressed in relation to the BOD_5, rather than the ultimate BOD, and the coefficient b is expressed in relation to the volatile solids, rather than to the living matter.

In the medium rate biological treatment of municipal wastewater (see below), the following values may be adopted as first approximations:

a' = 0.5 kg per kg of BOD_5

b' = 0.1 $kg.d^{-1}$ per kg of VS

a_m = 0.6 kg per kg of BOD_5

b = 0.05 $kg.d^{-1}$ per kg of VS.

FACTORS RELATING TO THE OPERATION OF A BIOLOGICAL REACTOR

In water treatment, a biological reactor may be defined by three basic parameters: the loading (F/M ratio and loading), the settleability of the sludge, and the sludge age.

F/M ratio and BOD or COD loading

The F/M ratio (or sludge loading) is the ratio of the mass of food (usually expressed in terms of BODs) entering the reactor per day and the sludge mass contained in the reactor:

$$F/M = \frac{Q.S_o}{X_t.V}$$

where,

Q : the daily flow,

S_o : the substrate concentration,

X_t : the concentration of mixed liquor suspended solids (MLSS),

V : the volume of the reactor.

It would be more logical to consider volatile solids X_v instead of total suspended solids X_t In this way, the F/M ratio would become F/M'.

$$F/M' = \frac{X_t}{X_v}.F/M$$

In line with common practice, the F/M ratio will be used to define the sludge loading.

This concept of F/M ratio is important in the case of activated sludge, since it determines:

- the purifying efficiency. Low rates correspond to a high purification efficiency and high rates correspond to lower purification efficiency;
- the production of excess biological sludge. With a low rate the endogenous respiration is greater than with a high rate due to limitation in the substrate; the production of biomass is therefore lower;
- the degree of stabilization of the excess sludge produced. Since forced endogenous respiration leads to a biomass with a high mineral level, low rate procedures are characterized by less fermentable excess sludge;
- the oxygen requirements associated with the removed pollution. Compared to high rate processes, the

amount of endogenous respiration with low rate processes results in higher levels of oxygen consumption relative to the pollution removed.

The various types of activated sludge treatment can be classified according to the F/M ratio at which they operate.

F/M ratio kg BOD_5/kg SS.d	Type of treatment
F/M < 0.15	Low rate (or extended aeration if F/M < 0.07)
0.15 < F/M < 0.4	Medium rate
0.4 < F/M	High rate

Another concept of loading is often used: volume loading. The loading is the food mass (usually referred to in terms of BOD_5) entering the unit per unit reactor volume per day:

$$C_v = \frac{Q\,S_o}{V}$$

usually expressed in kg $BOD_5/m^3 \cdot d$.

Sludge age

The sludge age A is the ratio between the mass of sludge present in the reactor and the daily mass of excess sludge extracted from the unit.

The daily production of excess biological sludge is given. If X_v refers to the concentration of volatile solids in the aeration tank and V is the volume of the tank, then:

$$A = \frac{X_v.V}{a_m.BOD_5 \text{ removed} - b.X_v.V}$$

If E is the efficiency of BOD_5 removal, the following simplified equation is obtained:

$$A = \frac{1}{a_m.E.F/M' - b}$$

Thus, sludge age is inversely proportional to the F/M ratio. This concept of sludge age is particularly important because it expresses the physiological scare of the microorganisms. The respiratory coefficients a' and b' cited earlier are closely correlated to sludge age. Moreover, the sludge age indicates the presence or absence of nitrifying bacteria.

Settleability

The efficient operation of an activated sludge facility depends on the correct operation of both the aeration rank and the clarifier. For the clarifier to efficiently separate the biomass from the treated water, the biomass must be properly flocculated.

Under certain conditions, the microorganisms agglomerate in floes. This is called *bioflocculation.*

During the phase of exponential growth, the bacteria remain dispersed throughout the culture. As soon as the declining growth phase begins, they agglomerate in brownish, jagged floccules chat are often several millimetres long. Seen under a microscope, the flocs frequently appear to be branched like the fingers of a glove, and the bacteria seem to be enveloped in a gelatinous substance.

The floc remains in the endogenous metabolic phase. However, by observing the correlations between the sludge age and the changes in the percentage of free organisms not associated with the floc, it is possible to determine that the minimum value lies in a range of four to nine days. After more than nine days, although the settleability remains adequate on the whole, a deflocculation process begins. The flocs become smaller, and increasing numbers of small particles escape (pin- head flocs).

Conversely, in less than four days, the very hydrophilic floc settles poorly and the number of free microorganisms increases very rapidly.

Bioflocculation is a complex phenomenon. For the time being, is has been firmly established that:

- it is controlled by the physiological scare of the cells;
- it is not unique to a single species, but is a fairly widespread phenomenon among common microflora;
- the basic effect is linked to the excretion of polymers, among which the polysaccharides play a special role.

A simple and practical way of assessing the ability of a sludge to settle is to determine its SVI.

Aeration

Oxygen is introduced into water by bringing the water into intimate contact with air. At the interface, the monomolecular boundary layer is saturated with oxygen as soon as it is formed. At the same time, the gas scares diffusing toward the deeper layers of water.

The quantity of oxygen diffused per unit of time is:

$$\frac{dm}{dt} = K_L \, a \, \frac{dC}{dt}$$

K_L is known as the transfer coefficient.

On the basis of this equation, the oxygenation capacity (OX.CAP.) of an aeration system is defined as the quantity of oxygen expressed in g.m^{-3} supplied in one hour to pure water with a constant zero oxygen content, at a temperature of 10°C, and at atmospheric pressure of 760 mm mercury.

The following equation allows the value of OX.CAP. to be calculated:

$$\frac{dC}{dt} = 11.25 \, \frac{1}{t_1 - t_0} \log \frac{C_s - C_o}{C_s - C_t} \sqrt{\frac{k_{10}}{k_t}}$$

where,

C_s : the O_2 concentration at saturation in g.m^{-3}

C_o : the O_2 concentration at the beginning of the experiment, in $g.m^{-3}$

C_t : the O_2 concentration at time t

11.25 : the saturation concentration of O_2 at 10°C in $g.m^{-3}$

t_1-t_o : the observation period in hours

dC/dt : the oxygenation capacity in $g/m^3.h$

k_{10} and k_t : diffusion coefficients at 10°C and t°C

Taking C_s - C_o = D_o and C_s - C_t = D_t as O_2 deficits at the start and at time t, the result is:

$$\text{OX.CAP.} = 11.25 \times 2.3 \frac{1}{t_1 - t_0} \log \frac{D_o}{D_t} \sqrt{\frac{k_{10}}{k_t}}$$

The relation between D_o and D_t plotted on logarithmic coordinates gives a straight line as a function of time, the slope of which, tan α, defines the oxygen dissolution rate:

$$\text{OX.CAP.} = 25.9 \tan \alpha \sqrt{\frac{k_{10}}{k_t}}$$

The quantity of oxygen introduced depends on:

- the value of the interfaces between the air and water, and their renewal;
- the oxygen concentration gradient between the air and water;
- the time available for oxygen diffusion.

There are, however, physical and technical limits to the optimal values for these conditions.

The size of the bubbles is an important parameter. However, they do have a lower limit, as the air bubble escaping from an orifice under water has a diameter much larger than that of the pore. In practice, the bubbles formed by porous aeration devices have a diameter of about one millimetre. Smaller bubbles can only be obtained by air release from air-saturated water (a process used for flotation).

All other factors being equal, the oxygen transfer coefficient K_L depends on the nature of the water (clean water, wastewater containing suspended or dissolved solids, the presence of surface-active agents), the aeration system used, and the geometry of the reactor.

In general, aeration systems are compared on the basis of their oxygenation capacity per m^3 of pure water per hour. The specific oxygen transfer capacity of a system can also be expressed in terms of the oxygen supplied per kWh.

Nitrification and Denitrification

In municipal wastewater as well as in the many types of industrial wastewater, nitrogen is present mainly in organic and in ammonia forms. In units designed to remove nitrogen, all the reactions can be illustrated in the following diagram:

raw water
$NH_4^+ \rightarrow NH_4^+$ ($\rightarrow NO_2^- \rightarrow NO_3^-$) ($NO_2^- \rightarrow N_2$)

Nitrification Denitrification

Ammonification

Organic N
raw water $\rightarrow$ assimilation = nitrogen removed in excess sludge (bacterial synthesis).

In the treatment of drinking water, the assimilation phase is insignificant.

Thus, the biological removal of nitrogen involves four main reactions.

Ammonification

Ammonification is the transformation of organic nitrogen into ammonia nitrogen. The ammonification rate depends essentially on the concentration of ammonia nitrogen. A Monod-type inhibition constant is used. Thus, if r_x is the ammonification rate, then:

$$r_x = \frac{r_x\left(N-NH_4=0\right).K_n}{K_n+N^*}$$

where :

N^* : the ammonia nitrogen concentration

K_n : experimentally determined as equal to 10 $mg.l^{-1}$

r_x, $(N\text{-}NH_4 = O)$ = 1.5 mg N per g of VS and per h.

In view of the nature of organic nitrogen and plant operation factors and, more particularly, the retention time of the water in the facility, most organic nitrogen will invariably undergo ammonification without problem.

Assimilation

Assimilation is the use of part of the ammonia nitrogen, and possibly the organic nitrogen, for bacterial synthesis. Assimilation can play an important role in removing nitrogen from certain types of industrial wastewater. However, in many cases and particularly in municipal wastewater, assimilation alone is not sufficient to remove nitrogen because the quantities present in the wastewater to be treated are much higher than those that can be assimilated for synthesis.

Nitrification

Nitrification is the process in which ammonia nitrogen is oxidized into nitrite and then into nitrate. Nitrification takes place in two stages and is brought about by autotrophic microorganisms. The process involves:

- the oxidation of NH_4^+ to NO_2^+, which is basically the work of bacteria known as *Nitrosomonas;*

- the oxidation of NO_2^- to NO_3^-. The bacteria responsible for this second reaction belong mainly to the genus Nitrobacter.

The entire simplified reaction may be represented as:

$$NH_4^+ + 2O_2 \rightarrow NO_3^- + 2H^+ + H_2O$$

Nitrification occurs only when nitrogen is initially present in its ammonium form. The transformation rate of ammonium into nitrate in an activated sludge process is of the order of 3 mg of N-NH_4^+ oxidized to NO_3^- per g of VS and per h. The value is, therefore, relatively high.

The generation time of bacteria in nitrification is often the factor that determines the design of equipment used for nitrification purposes. For these bacteria, the μ_{max} values are as follows:

- Nitrobacter: $\mu_{max} = 0.03\ h^{-1}$
- Nitrosomonas: $\mu_{max} = 0.08\ h^{-1}$

This order may be inverted under certain conditions, for instance when treating cold drinking water with low ammonium concentration, but the limiting factor is usually the bacteria responsible for the oxidation of ammonium into nitrite.

Because the growth rate of the heterotrophic bacteria responsible for the oxidation of carbonaceous pollution is higher than that of the autotrophic nitrifying bacteria, the sludge age in the selected purification system has a determining effect on nitrification. Under pH conditions of between 7.2 and 8, the minimum sludge age (in days) to start the nitrification in an activated sludge system is related to temperature as follows:

$$A_{minimum} = 6.5\ (0.914)^{T-20}$$

where, T is the temperature in °C.

The sludge age depends greatly on the temperature, and that nitrification at temperatures below 12 or 13°C leads to the adoption of low rate processes.

It is extremely risky to start the nitrification process at temperature below 8°C. However, if the nitrifying flora can be developed beforehand and introduced at normal temperatures, the nitrification process can continue at very low temperatures, resulting in a reduced oxidation efficiency of ammonia nitrogen. Nitrification requires additional oxygen. It is necessary to supply 4.6 mg of oxygen per mg of $N(NO_3^-)$ produced; or more precisely, 4.2 mg, taking into account the oxygen released by HCO_3. The nitrifying bacteria tolerate, however, periods of insufficient oxygen relatively well.

One point to be emphasized is the destruction of alkalinity. The growth rate of nitrifying bacteria falls substantially when the pH drops below 7.2. The alkalinity of raw wastewater is an important consideration, since the nitrification reaction produces H^+ ions. Calculations and experience show that 7.2 mg of alkalinity expressed as $CaCO_3$ (0.72 Fr. deg.) are required to neutralize the H^+ ions that are produced by the oxidation of 1 mg of $N\text{-}NH_4^+$.

This explains why, in certain soft water areas, when no correction is made the nitrification process results in a pH that is too low for the complete oxidation of ammonia.

Lastly, nitrifying bacteria are sensitive to many factors in their environment. The presence of certain organic or inorganic compounds may modify or even inhibit the growth of nitrifying bacteria, particularly in certain industrial wastewaters.

Denitrification

Denitrification is a process where certain bacteria reduce nitric nitrogen to a lower oxidation state.

These bacteria may be autotrophic, but their activity is weak, which is why heterotrophic bacteria are generally used.

This reduction takes place in the form of several reactions, which may be expressed as follows:

$$NO_3^- \rightarrow NO_2^- \rightarrow NO \rightarrow N_2O \rightarrow N_2$$

In the treatment of wastewater, the following reaction is acceptable:

$$NO_3^- + 6H^+ + 5e^- \rightarrow 0.5\ N_2 + 3H_2O$$

The preferred source of electrons is organic carbon, or the bacterial mass itself.

Four essential factors influence the denitrification rate: temperature, dissolved oxygen, pH, and the source of organic carbon.

Temperature

The influence of the temperature may be expressed by the following equation:

$$r_T = r_{20}{}^oC.\theta^{(T-20)}$$

where, r_x and $r_{20}{}^oc$ are the denitrification rates at temperatures T_x and 20°C expressed in mg of $N\text{-}NO_3^-$ reduced per mg of VS and per h. θ, which is determined experimentally, equals 1.116 for temperatures between 5 and 27°C

Dissolved Oxygen

The presence of oxygen inhibits denitrification. Denitrifying bacteria draw their energy from the energy released in the transfer of electrons from organic compounds to O_2, NO_2^- or NO_3^-. If these three elements are available to accept electrons, then the preferred recipient is clearly the element that yields the greater quantity of energy per unit of oxidized organic matter. Since more energy is yielded with oxygen than with the other elements, electrons are transferred preferentially to oxygen.

In reality, it is difficult to fix concentration limits for O_2, above which denitrification is inhibited, since values depend on how large the floc is and how easily oxygen transfers to the bacteria. These values also depend on the nature and concentration of organic carbon. However, O_2 concentrations higher than 0.5 $mg.1^{-1}$ are known to be highly inhibiting. Another approach is to measure the redox potential instead of the dissolved oxygen. Even though the figures observed frequently vary, it seems that denitrification is not satisfactory unless the redox potential is less than 220 mV.

pH

The optimum pH lies between 7 and 8.2.

Source of carbonaceous substrate

The nature and concentration of the carbonaceous substrate both have an effect. The following table gives the kinetics of nitrate reduction for several substrates expressed in mg of N-NO_3^- reduced per g of VS and per hour, at a temperature of 20°C. These values of r_{20}°c were obtained with denitrifying activated sludge.

Ethanol	5.1
Acetate	4.9
Propionate	5.1
Methanol	2.5
Butyrate	5.1
Municipal wastewater	3.3
Endogenous respiration	1.5

Denitrification using the organic carbon present in municipal effluent thus achieves wholly acceptable kinetics as compared with the kinetics obtained with easily biodegradable products.

However, in some cases the quantities of organic carbon introduced by an effluent may be insufficient to obtain advanced denitrification. It should be noted that the use of organic carbon released by endogenous respiration results in weak kinetics. In practice, this requires structures that are very large in volume, which explains why this technique is not often implemented.

Biological Phosphate Removal

Many research projects have studied the possibility of removing phosphates biologically, *i.e.*, without introducing a reagent and practically without producing additional excess sludge. These studies really began in the mid 1960s with the work of Shapiro and Levin. They observed that non-aerated activated sludge released phosphorus and then reabsorbed it as soon as the oxygen concentration increased.

The principle of biological phosphate removal consists in accumulating phosphorus in a biomass. This accumulation may be due either to the chemical precipitation of inorganic phosphorus in the vicinity of bacteria under particular microenvironmental conditions, or to an accumulation of phosphorus by the microorganisms themselves, or to a combination of the two.

Extracellular Precipitation of Inorganic Phosphorus

The main causes of the formation of these precipitates would be an increase in the pH or an increase in the concentration of precipitating ions. Several observations confirm the possibility of these phenomena. When a bacterial growth is subjected to anaerobic conditions, its level of extracellular calcium falls, and it discharges phosphorus, potassium, and magnesium: potassium and magnesium ions stabilize intracellular polyphosphate. The release of phosphate ions appears to cause a drop in the calcium concentration, which supports the hypothesis of precipitation.

In the absence of oxygen, changes in pH due to denitrification and the acid fermentation of organic products could cumulate with the effects of increased phosphorus concentration and either accentuate or lessen the consequences.

Intracellular Accumulation of Polyphosphates by Microorganisms in Activated Sludge

In addition to extracellular precipitation, which is highly variable and difficult to assess and control, it has now been determined that bacteria accumulating polyphosphates also play a key role (poly-P). The phenomenon of polyphosphate storage has often been observed in microbiology, especially in cases where an imbalance in the nutrient medium impedes the synthesis of nucleic acids. The accumulated polyphosphates may serve either as an energy reserve comparable to the phosphate chain in the ATP/ADP system, or as a phosphorus reserve. Nuclear magnetic resonance (NMR) analyses on phosphorus-removing sludge have confirmed the presence of substantial quantities of poly-P, which is invariably found in the form of "volutin" granules.

Factors Involved in Biological Phosphate Removal

The biological removal of phosphates requires alternating anaerobic and aerobic sequences. The aim of alternating these sequences is to modify the enzymatic equilibrium regulating the poly-P synthesis in the anaerobic phase.

The Anaerobic Phase

Acetate-producing bacteria, facultative anaerobes, use the available organic carbon, existing for example in raw water, to produce acetate. Aeromonas are the main organisms responsible for this anaerobic acidogenesis. It should be noted that the presence of nitrates in this anaerobic phase prevents the production of acetate. This point is explained by the denitrifying capabilities of Aeromonas, which do not use their fermentative metabolism as long as there are nitrates in the medium to act as final electron acceptors.

The acetate produced is reused by the bacteria of the Acinetobacter/Moraxella group. These bacteria are strictly aerobic and can only use a limited range of substrates. They consume acetate, ethanol, lactate, citrate and several amino acids, but they cannot use sugars or volatile fatty acids with a molecular weight higher than that of propionic acid.

The acetate used by the Acinetobacter/Moraxella is stored in situ in the form of PHB (polyhydroxybutyrate). The energy required for this storage comes from the hydrolysis of poly-P, which explains the accelerated release of phosphate into the medium. In this process, acidogenesis is the limiting step, which explains why free acetate is not found in anaerobic zones.

Aerobic Phase

The Acinetobacter/Moraxella locate the electron acceptors for their metabolism (NO_3, oxygen). The PHB is then used as organic substrate for their growth and the replenishment of their reserve of poly-P through the reabsorption of interstitial phosphorus. The quantities reabsorbed exceed the amount released during the anaerobic phase. Thus, in a succession of anaerobic-aerobic phases, phosphorus is gradually accumulated in these microorganisms in quantities reaching 10 to 11% of dry weight. In a phosphorus-removing activated sludge in which other bacteria live, this value could reach 7% of dry weight.

It should be noted that in the aerobic phase, other heterotrophic organisms that do not accumulate poly-P have at their disposal only a fraction of the organic matter, which is, moreover, the least biodegradable.

The common denominator in all these processes using the biological removal of phosphates is thus the alternating of an anaerobic phase, in which the biomass comes into contact with the organic carbon in untreated water, with an aerobic phase in which the previously released phosphorus is reassimilated.

These, processes fall into two general categories:

- processes in which no chemical reagent is added; in these systems the phosphorus is "biologically" stored in the sludge and removed with the excess sludge. The efficiency of phosphate removal thus depends wholly on the phosphorus content of the sludge and on the production of excess sludge;
- processes in which the phosphate removal is a combined biological and physical-chemical process. The biologically accumulated phosphorus in the sludge is released in a small volume of water. A high concentration of phosphorus is thus obtained in the interstitial liquid, to which chemical reagents are added.

Aerobic Stabilization

This process, which is designed to reduce the level of OM in sludge, is also often referred to as "aerobic digestion". In this method, the extended aeration of the sludge is used to stimulate the growth of aerobic microorganisms beyond the period of cell synthesis and depletion of the substrate until the stage of auto-oxidation is reached. This is the mechanism of endogenous respiration.

The cellular matter (represented as $C_5H_7NO_2$) is transformed in the following reaction :

$$C_5H_7NO_2 + 5O_2 \rightarrow 5CO_2 + NH_3 + 2H_2O$$

The ammonium may then be oxidized, resulting in the following overall reaction:

$$C_5H_7NO_2 + 7O_2 \rightarrow 5CO_2 + NO_3^- + H^+ + 3H_2O$$

The pH may then drop substantially if the alkalinity of the wastewater is insufficient.

Aerobic stabilization is used most often with activated sludge, but may also be applied to sludge produced in plain, primary settling, as well as to biofilms in attached growth reactors. Thus, the reduction rates of OM (E%) obtained, vary widely. E% may generally take the form E = p.log t + q, where t is the stabilization time. The reduction rate in the case of activated sludge with an age of five days, obtained from nonsettled municipal wastewater.

Oxygen requirements for a stabilization time of 15 days at 15°C vary from 0.1 to 0.15 kg of O_2 per day and per kg of VS, depending on the source of the sludge. This oxygen consumption also depends on the temperature.

Koers and Mavinic consider that the product (temperature °C x stabilization time) can be used to calculate the reduction rate of organic matter for sludges of the same type.

Various quantitative criteria have been proposed to define a sufficiently stabilized sludge :

- the respiration of the sludge: 0.1 kg 0.2 per kg of OM per day, at 20°C;
- weight loss of less than 10% after 120 h of aeration at 20°C in a medium with an O_2 level of 2 $mg.l^{-1}$.

Another proposed method is to assess the amount of volatile acids formed and the amount of nitrogen ammonified per gramme of **OM** maintained under anaerobic conditions.

The term **thermophilic aerobic digestion** refers to the stabilization process that, by limiting hear losses, benefits from the exothermic nature of the OM oxidation reaction to heat the sludge to more than 50°C.

ATTACHED GROWTH

The Biological Film

Most microorganisms are able to grow on the surface of a solid when organic compounds, mineral salts and oxygen are available. They are anchored by means of an exopolymer-based gelatinous material produced by the bacteria, inside which the bacteria can, to some extent, move about. The colonization of the solid matter begins in selected areas, whence the biofilm develops continuously until the entire surface of the support is covered with a monocellular layer. From this moment on, growth is carried on by the production of new cells covering the first layer.

The oxygen and nutrients carried in the water to be treated diffuse throughout the biofilm until the deepest cellular accumulations are no longer affected by the oxygen and nutrients.

After some time, stratification occurs with an aerobic layer, where the oxygen is diffused, on top of a deeper anaerobic layer in which there is no oxygen. The thickness of these two layers varies according to the type of reactor and support.

The use of biofilm methods for treating water shows that:

- the rate at which the substrate is used Stabilizes at a constant value when the biofilm is deep enough for oxygen to become a limiting factor in the deepest layers. The "active" depth may be about 300 - 400μm;
- the bacteria attached to a support usually display higher specific activity than those observed in suspended growth.

Trickling Fiber (Bacteria Beds)

The operating principle of a trickling filter (sometimes referred to as a biofilter or as a percolating filter) consists in allowing the water to be treated to trickle onto a mass of material with a specific surface area of between 50 and 200 $m^2.m^{-3}$ and supporting a film of purifying micro-organisms. Depending on the type of material used, there are two basic categories of trickling filter:

- trickling filters with traditional fill. The fill can be pozzuolana, blast furnace coke, or crushed siliceous rocks with a void ratio of about 50%. This type of biological filter is predominantly used for municipal wastewater;
- trickling filters with plastic fill, often used in the treatment of industrial wastewater. These plastic materials, which are used in a random or ordered fashion, have void ratios above 90%.

If the substrate fails to reach them, the microorganisms in the anaerobic layer die, then undergo autolysis. In this way, their cellular contents become available to other facultative aerobic or anaerobic microorganisms. When all of the substrate reserves are truly exhausted, the lysis of the remaining cells causes the biofilm to become locally detached from the surface. This surface then becomes available for a new colonization. The action of water currents on the surface of the biofilm can bring about or contribute to its detachment (sloughing). Whatever the material, all trickling filters operate on the same principles. The filter is aerated, in most cases by natural draught, but occasionally by forced countercurrent ventilation.

Basic Equations

The BOD removed by a trickling filter depends on the nature of the water to be treated, the hydraulic load, the temperature, and the type of fill material.

The mathematical formulation is based on the assumption that the microorganisms in the trickling filter are in the declining growth phase:

$$\frac{S_f}{S_o} = \exp.(-k_1 t)$$

where,

S_f : BOD_5 of the clarified effluent

S_o : soluble BOD_5 from the feed to the filter

t : average retention time of water in the filter

k_1 : a constant depending on:

- water temperature
- the type of fill
- the nature of the wastewater to be created.

The median value of t is expressed as:

$$t = k_2 . \frac{H}{Q^n}$$

where,

H : filter depth

Q : flow per unit of horizontal surface

k_2 and n : constants

Hence,

$$\frac{S_f}{S_o} = \exp.\left(-k_1.k_2.\frac{H}{Q^n}\right)$$

Trickling Filters with Traditional Fill

With traditional fill materials and a bed depth of 2 m, purification efficiency is relatively low (66%) when the loading (expressed in kg of BOD_5 per m^3 of material per day) is high. In this case, efficiency can be increased by recirculating the filter effluent back to the filter, thus diluting the feed water.

Empirical equations have been developed for domestic wastewaters. For example, Rankin gives the following equations, based on a maximum hydraulic load (including recirculation) of 1.13 $m^3/m^2.h$:

- single-stage trickling filter :

$$S_f = \frac{S_o}{2r + 3}$$

- two-stage trickling filter:

$$S_{f1} = 0.5\ S_o$$

$$S_{f2} = \frac{S_{f1}}{2\,r + 2}$$

where, r = recirculation rate.

Depending on the loading used, a distinction is made between low-rate and high-rate filters, which have the following performance ratings with municipal wastewaters:

Loading	Low rate	High rate
BOD $kg/m^3.d$	0.08 to 0.15	0.7 to 0.8
Hydraulic loading $m^3/m^2.h$	< 0.4	> 0.7

In high rate filters, which normally require recirculation, the hydraulic loading is sufficient to homogenize the bacterial flora at the various levels. Self-cleaning of the material, which then retains only a thin active film, encourages rapid exchanges and relieves the trickling filter of the task of breaking down the cellular material that develops. This process of mineralization (stabilization) occurs in other sections of the facility, such as the anaerobic digester, which means that a clarifier must be used at the filter outlet to collect the settled matter for transfer to the sludge treatment facilities.

In a high rate filter, the activity of predatory agents is limited. On the other hand, in a low rate filter there is no continuous washing of the sludge, which tends to build up in the contact mass. The action of predatory agents is essential, and this, together with endogenous respiration of the bacteria, limits excessive growth of the film.

Nitrification may take place if the organic carbon load applied is sufficiently low. The municipal wastewater settled at 15°C (G. Martin), gives the TKN removal efficiency as a function of the loading in a trickling filter with pozzuolana packing.

Unlike in activated sludge, nitrification is not an "all or nothing" process, because the microorganisms are stratified throughout the entire depth. The top portion contains primarily heterotrophs, while the bottom part is mostly made up of autotrophs. In contrast to activated sludge, the limiting factor is not the sludge age, but the kinetics of nitrification.

Owing to the depth of the zoogloea attached to the support and the fact that oxygenation is often insufficient, trickling filters frequently enable partial denitrification to take place. This denitrification is improved by recycling and it essentially occurs in the upper layers of the trickling filter, where the organic carbon from the raw water is available.

In a low rate filter, the sludge is strongly mineralized and can be discharged into the outlet without final clarification if periodic discharges of sludge into the final effluent are permissible.

Because of the frequent risk of clogging, the large fly population they encourage and their high cost, low rate filters are little used despite their high efficiency (95% for BOD_5). High rate filters, with recirculation, are more commonly used for removing carbonaceous pollution.

Recirculation

Recirculation has several advantages:

- self-cleaning of the trickling filter,
- seeding of the settled effluent,
- diluting of the high BOD wastewater.

The most frequently used recirculation system. The secondary sludge is continuously recycled. Since the recirculation flow is drawn from the bottom of the clarifier, the surface area of the clarifier can be designed exclusively for a rising velocity corresponding to the flow Q to be treated. On the other hand, the primary settling tank must be designed to take Q(l + r), where r is the recycle rate.

The fill materials used for trickling filters must be dean and non-friable. Particle size must be regular and between 40 and 80 mm. The void ratio is about 0.5, which means that, allowing for the biological film, the void left free for aeration is limited to about 0.15. Whatever the traditional material used, the risk of clogging by coarse suspended solids from the raw water requires the construction of a primary settling tank upstream of the filter.

Trickling Filters with Plastic Fill

Trickling filters with traditional fill are little used for the treatment of high BOD industrial wastewater flows from sources such as the agrifood industries, because of the risk of clogging and the excessive proliferation of filamentous biological films.

Plastic fill considerably reduces these risks and provides a higher oxygen transfer coefficient. The applications of plastic packing filters are very different from those with traditional fill for the following reasons:

- they may be operated with high BOD_5 loadings of between 1 and 5 kg/m^3.d, or even higher;

- since plastic is more expensive than traditional fill, the aim is to use plastic-media high rare trickling filters. Under these conditions, the BOD_5 removal is too low to yield an effluent complying with the standards usually in force, since it fluctuates between 50 and 80% depending on the type of flow treated and the loading chosen. For this reason, treatment by a trickling filter with plastic fill is often followed by a stage of conventional treatment, such as activated sludge.

Random packings seem to become clogged more readily than ordered materials, and their use is generally limited to low pollution flows free of suspended solids, fibres or greases. They may also be used in nitrifying filters downstream of an activated sludge stage. Thus, the nitrification rate for industrial wastewater is approximately 100 g of N-NH4 per m^3 and per day at 20°C and 50 g at 15°C.

Granular Beds

The activity of a bacterial growth depends in particular on its exchange surface with the substrate and oxygen. In activated sludge this surface is restricted by the flocculated state of the microorganisms. This flocculated state is indispensable for proper separation of sludge and treated water in the final separation tank. The larger the floc, the more slowly the substrate and oxygen are diffused toward the microorganisms inside.

In trickling filters, the developed surface of the support material, related to m^3 of reactor, remains small and the perfect distribution of water to be treated across the entire surface of the biofilm is difficult on an industrial scale.

When microorganisms are attached to granular supports with an effective size of less than 4 or 5 mm, the supports offer a specific developed surface area, and consequently an exchange surface, that is much larger than the surface offered by other processes. For example, Biolite with an ES of 2.7 mm, provides an exchange surface of 700 $m^2.m^{-3}$. A turbulent bed, using a support with an ES of 375 μm, expanded by 100%, develops a surface exchange area of 6500 $m^2.m^{-3}$ of reactor, compared with 100 or 200 $m^2.m^{-3}$ in a trickling filter with plastic fill.

It is thus possible to operate at high BOD loadings and still obtain, in a single stage, purification efficiencies that produce municipal wastewater effluent meeting usual discharge standards.

These granular-media bioreactors may be divided into two main groups:

- those which perform biological purification and retain the SS initially present in the raw wastewater as well as the excess sludge produced. These are fixed granular beds, also known as biofilters;
- those in which the bioreactor performs only biological purification, with SS being removed by a unit situated downstream. In this case, the exchange surface for microorganisms and substrates is optimized by using very fine granular materials that are kept in motion. These units are known as moving *granular beds.*

Fixed bed reactors may operate in upflow or downflow, according to the direction of the water flow chosen. Depending on the relative directions of the water to be treated and the oxygenating gas, a distinction may also be made between cocurrent and countercurrent reactors.

The granular support material is periodically *regenerated* by washing in the reactor itself. Note that, in theory, the principle of purification applied in fixed granular beds can be extended to include floating granular beds made up of low-density beads.

Moving beds can be:

- **fluidized**, where oxygen is dissolved in the recirculated effluent. However, controlling the growth of the biological film requires an extraction, washing and material recycling system. The loadings used are limited by oxygenation capacity.
- turbulent, where air is directly introduced into the reactor. A settling rank downstream is essential. The purification efficiency is limited.

Other Systems

Biological Discs

This method, which is also known as the Rotating Biological Contactor (RBC), goes back to the 19th century, to the work of Weigrand on the purifying capabilities of water mill wheels.

The biomass is attached to discs that turn around a horizontal axis and are partially bathed in the water to be treated. Rotation brings the biomass alternately in contact with the water to be treated and the oxygen in the air.

An electric motor usually provides the energy for the discs to rotate. Several methods have been designed to aid rotation and oxygenation by blowing additional air into pockets attached to some of the discs. The discs, which are made of polystyrene, PVC, or corrugated polyethylene sheets, are 2 to 3 m in diameter.

They are spaced 2 to 3 cm apart and turn at 1 to 2rpm.

A clarifier, designed for rising velocities of up to $2 m.h^{-1}$, retains the excess sludge.

The absence of any stirring in the aeration tank:

- necessitates the presence of a primary settling tank.
- prohibits the recirculation of sludge after the clarifier.

These systems are often made up of several disc stages, the first of which remove organic carbon, and the last of which perform nitrification. Loadings are expressed in g BOD_5 per m^2 of disc surface per day. Loadings rarely exceed 25 to 30 $g/m^2.d$. With considerably lower loadings, nitrification is possible but the system is highly temperature-sensitive.

The advantage of this method is that it consumes little electrical energy (2 to 4 $W.m^{-2}$ of disc), but widespread use has been hindered by:

- the need to stabilize primary and biological sludge;
- the great difficulty in obtaining a treated municipal wastewater effluent with less than 40-45 $mg.l^{-1}$ BOD_5 without making heavy additional investments;
- the need to cover the discs to protect them against harsh weather.

Submerged Contact Structures

This method consists in submerging in a rank of activated sludge a fixed or floating structure on which an additional biomass has developed, which is not required to pass through the clarifier. Thus, it is theoretically possible to improve the performance of a biological purification facility without enlarging the clarifier, which may be limited by the solids loading applied.

Another more promising application involves nitrification ranks that are fitted with these contact structures and placed downstream of a conventional facility removing carbonaceous pollution. This system can be compared with a method that is widely used in small facilities in Japan, where a final aeration tank fitted with honeycomb modules is located downstream of the small plane with no final settling rank.

The criteria determining the choice of contact structure are:

- the specific surface area. The growth of the biomass concentration depends directly on the surface area. Processes may be grouped on the basis of this criterion, which determines all the subsequent technological options;
- sensitivity to clogging and the possibilities of cleaning;
- the resistance to wear and tear;
- the material and installation costs.

The processes differ mainly in the type of materials used:

- flat materials. These processes use plastic fill similar to that used in trickling filters. The BOD loadings applied remain below 2 kg/m^3.d. The increase in the level of sludge is about 20 to 40% as compared to traditional activated sludge;
- filiform materials. The threads employed may be used in two ways:
- by direct implantation of threads arranged in various ways (loops, clusters, etc.), mainly using the "ring-lace" technique.
- by using 2 to 3 cm edge cubes made of polyurethane mesh.

  The major drawback of this method lies in the especially high risks of clogging and agglomeration, particularly with waters containing fibres, greases, etc.
- floating materials.

AEROBIC BIOFILTERS

Aerobic biofilters (also known as 'trickling filters' or 'percolating filters') have been in use for the treatment of organic wastewaters since the first decade of this century. They were originally developed from the practice of sewage treatment by application to land, or 'sewage farming', as it was known in the nineteenth century.

Attempts to improve the efficiency of land treatment led to the discovery in about 1888 that settled sewage could be purified by passing it through an artificial bed of coarse porous medium at about 10 times the application rate for conventional land treatment. Following this discovery, percolating filters were developed fairly rapidly and by 1908 this method had become the standard method of accelerated sewage purification. Today, aerobic biofilters compete with the activated sludge process for the purification of dilute organic wastewaters, while high-rate biofilters are sometimes used for the partial treatment of strong organic wastewaters.

Aerobic biofilters may be classified as biological reactors of the attached film/static medium type. The medium most widely used is natural stone (50-100 mm size), while a variety of purpose-made plastic media is also available.

MODE OF ACTION

The mechanism of removal of organics is similar to that of the activated sludge process. The greater portion of the liquid applied to the surface of the filter passes rapidly through and the remainder trickles slowly over the surface of the slime growth. The removal of organic colloids in suspension and dissolved substances occurs by 'biosorption' and coagulation from that portion of the flow that passes through rapidly, and by the usual processes of synthesis and respiration from the part of the flow with long residence time. This residence time is primarily related to the hydraulic loading, so it seems reasonable that the greater the hydraulic loading the more the process will depend upon biosorption and the less it will depend upon synthesis and respiration, other things being equal.

The action of a filter depends on the metabolic activity of zoogleal or filamentous bacteria or of fungi. These colonize the extensive surfaces of the support medium and form the basis of the 'film' which also contains a population of protozoa as well as amorphous solids derived from the waste. Algal growths may also be present in the upper regions of the filter exposed to light. A population of macroorganisms is also usually found associated with the film. The composition of the microbial population depends on the nature of the waste, its strength, the rate at which it is applied and the method of operation of the filter. Therefore the film may range in character from a tenuous bacterial slime in a filter receiving a waste of low organic content, to a thick mass of fungal mycelium in a filter treating a strong sewage containing certain types of industrial wastes.

The reduction in concentration of polluting matter occurs most rapidly in the upper regions of the filter. In a conventional filter treating sewage, about 90% of the biodegradable matter may be removed in the upper 0.6 m of the bed. The net removal rate of organic matter is a function of the immediate removal of the readily biodegradable fraction and the release of products of endogenous metabolism of the biofilm. When ammonia is present in the waste, as in the case of sewage, its removal by microbial nitrification occurs largely in the middle and lower regions of the filter. The zones of carbonaceous oxidation and nitrification are not sharply defined, but conditions for the growth of nitrifying bacteria tend to be more favourable in the lower parts of the filter where the content of oxidizable organic matter in the waste has been reduced and where the oxygen concentration in the film is therefore higher (Bruce, 1969).

Mechanical filtration is not an important part of the purification mechanism, so that the term 'filter' is a misnomer persisting from the early days when the nature of the process was misunderstood. Although solids present in the waste may be physically trapped by the medium or the film matrix, it is essential for the satisfactory operation of conventional filters to remove as much coarse suspended matter as possible by primary sedimentation.

As a result of the growth of bacterial cells and of the deposition of coagulated solids from the waste, the quantity of film within the filter tends to increase. The growth rate is dependent on the concentration of degradable organic matter in the waste and on the application rate. The growth of the film tends to be greatest in the upper regions of the filter where the largest proportion of the substrate is removed. If the growth rate of the film exceeds its removal rate by other mechanisms, the interstices of the medium will ultimately become choked, impeding the flow through the bed and restricting ventilation. In the extreme, the surface of the filter may become ponded. This clogging of the medium with accumulated film is the most serious problem in the operation of conventional filter beds and it is the factor that limits the maximum rate at which wastes can be treated efficiently by such plants.

ECOLOGICAL ASPECTS (BRUCE, 1969)

The biological film or slime layer is inhabited by an interdependent microbial population, including bacteria, fungi, protozoa and a variety of macroinvertebrates. Algae are restricted to the surface of the bed, where light is available. They play a minor role in the purification process, but excessive growth can cause blinding of the surface.

The thickness of the film, which can be maintained in an aerobic condition by diffusion of atmospheric oxygen through the film surface, is rather limited. Estimates of the aerobic zone in an actively respiring film vary between 0.06 mm and 2 mm, while in deeper regions of the film anaerobic conditions prevail.

Macroinvertebrates include a variety of worms, insect larvae and snails. They serve a useful role in film disintegration, especially in low-rate filters, where the scouring effect of the liquid flow is rather limited.

One disadvantage of the presence of fly larvae in a filter is the emergence of adult flies, particularly *Psychoda* and *Anisopus*, which may cause a serious nuisance during the warmer months. Various methods (Tomlinson and Jenkins, 1947) for controlling the fly population have been used including treatment of the filter with lime, creosote, bleaching powder and salt. The most effective method of artificial control is to use an efficient insecticide in amounts sufficient to kill the flies and their larvae without affecting the worms and other animals in the filter, and also without rendering the effluent toxic to fish. It has been suggested that the only satisfactory long-term solution to the problem lies in the biological suppression of the fly population by a dominant population of worms competing for the available food supply.

SEASONAL VARIATIONS

Variations in temperature have a greater influence on the behaviour and performance of percolating filters than on those of activated sludge plants. The effect of the decline in the temperature of sewage during the winter is to reduce the activity of the macroinvertebrates. The growth rate of the film may also be reduced with temperature but generally to a smaller extent, and a net accumulation of film usually results. This may cause partial or complete blocking of the medium. Fungal growths flourish and at the same time worms and fly larvae, which would be present in the surface layers during the summer, retreat into the depths of the bed. The unequal effect of low temperature on film growth and animal activity limits the rate at which sewage can be applied to a filter, if an effluent of good quality is to be achieved. The arrival of warmer weather during the spring gives rise to a fairly rapid depletion of accumulated film in the filter and a corresponding increase in the concentration of suspended matter in the effluent. This phenomenon has not been completely explained. It has been attributed to the increased activity of the fly and worm populations in response to the temperature and also to microbial lysis within the film. The findings (Bruce, 1971) of pilot plant studies on high-rate filters at the Water Research Centre, Stevenage, showed a very marked sensitivity to temperature, the minimum winter BOD removal rate being less than half the maximum summer BOD removal rate.

PROCESS DESIGN

Aerobic biofilters are categorized as low-rate or high-rate, depending on the applied hydraulic and organic loading rates. Loading parameter ranges for both categories are summarized in Table 1. The conventional low-rate filter, used in domestic wastewater treatment, is typically loaded at about 0.1 kg BOD_5 m^{-3} d^{-1}, the corresponding hydraulic loading being about 1 m^3 m^{-2} d^{-1}. Space requirements are thus considerably greater than required for the activated sludge process, even compared with the extended aeration version of the latter.

Table 1

Parameter	Low-rate filter	High-rate filter
Hydraulic loading (md^{-1})	1–4	10–40
Organic loading (kg $BOD_5m^{-3}d^{-1}$)	0.1–0.3	0.3–2.0
Depth (m)*	1.5–3.0	1.5–6.0
Recirculation	none	usual

Stone and similar solid aggregrate media filters are generally not more than 2 m in depth, while high-rate plastic medium filters may be up to 6 m deep (so-called biotowers).

In high-rate filters, the recirculation of filter effluent permits the use of higher organic and hydraulic loading rates.

Because of the biological complexity of filter systems, accurate modelling of reactor kinetics has proved to be difficult. However, a number of empirical equations, quantifying filter performance, have been proposed (Metcalf and Eddy Inc, 1991). The NRC (1946) equations are amongst the simplest and most useful for design purposes. They were derived from an extensive study of the operating records of trickling filter plants serving military installations in the United States. The formulae are applicable to single-stage and multi-stage stone media systems, with varying recirculation rates. For a single-stage filter:

$$E_1 = \frac{1}{1 + 0.44\sqrt{\frac{W}{VF}}} \qquad (1)$$

where E_1 is the fractional efficiency of BOD removal for the biofilter, inclusive of recirculation and sedimentation; W is the BOD_5 loading rate (kg d^{-1}); V is the filter volume (m^3), and F is a recirculation factor, calculated as follows:

$$F = \frac{1 + R}{(1 + R/10)^2} \qquad (2)$$

where, $R = Q_r/Q$; F is the average number of passes through the filter. The term R/10 takes into account the experimental observation that the removal of organics appears to decrease as the number of passes increases. For the two-stage filter, the suggested NRC equation is:

$$E_2 = \frac{1}{1 + \frac{0.44}{(1 - E_1)}\sqrt{\frac{W^1}{VF}}} \qquad (3)$$

where E_2 is the fractional efficiency of BOD_5 removal for the second stage, including recirculation and sedimentation; W^1 is the second-stage BOD_5 loading rate, kg d^{-1}.

The influence of temperature on high-rate filter performance (see the earlier comments on seasonal influences) should be borne in mind when applying the NRC formulae to local conditions. It is probable that the efficiency levels computed from the NRC equations would only be attainable under summer conditions in temperate climates.

Alternating double filtration (ADF) is a further variation in filter layout. It consists of two filters operating in series, each with its own sedimentation tank. At intervals from daily to weekly, the order of the filters is reversed. The relegation of one of the filters to a secondary role causes a depletion in accumulated film and hence controls clogging. Thus, while a single low-rate filter might be subject to excessive film growth if the applied BOD_5 concentration exceeds 500 mgl^{-1}, the ADF arrangement overcomes this difficulty.

DESIGN OF PHYSICAL FACILITIES

The main elements of an aerobic biofilter system are:

(1) an influent feed system by gravity or pump;

(2) an influent irrigation system;

(3) containment for the filter medium;

(4) provision for adequate ventilation of the filter medium; and

(5) secondary sedimentation and recirculation of effluent, where required.

Influent Feed

For a single-stage filter process without recirculation, the available site gradient will determine if influent pumping is necessary. The head loss through a filter bed may be taken as the bed depth plus an allowance for upstream and downstream hydraulic losses of, say, 0.5 m. Where such a natural ground fall is not available, the clarified effluent from primary sedimentation is discharged to a pump sump, from where it is pumped to the filter distribution system.

Where recirculation is being applied, recycled effluent is also discharged to this sump.

Distribution Systems

The standard method of clarified effluent irrigation on circular filters is by rotary distributor, which consists of two or more arms, mounted on a pivot at the centre of the filter and revolving in a horizontal plane, as shown in Figure Each arm is designed as a distribution manifold with discharge points so spaced as to secure a uniform irrigation of clarified wastewater over the entire bed area (requires a decreasing discharge nozzle spacing with distance from the centre of rotation).

The distributor may be driven by the jet reaction of the nozzle discharges or by a geared motor. The speed of rotation is selected to give an application interval of 4-12 mins, which has been found (Tomlinson and Hall, 1955) to give satisfactory process performance. A clearance of at least 0.15 m is usually allowed between the underside of the distribution arm and the top of the bed. Motor-driven distributors have the advantage of providing a constant rotational speed, while the speed of reaction-driven distributors is determined by the flow rate. In small-scale units of the latter type, the inflow rate may be syphon-controlled to secure a measure of uniformity in application rate and hence in the distributor rotational speed.

Rectangular filters are less frequently used. Wastewater irrigation systems for rectangular filters generally incorporate a travelling manifold, which traverses the filter length in both directions. A fixed manifold distribution system is normally used for wastewater irrigation on deep-bed filters, so-called biotowers. These units incorporate self-supporting plastic media and may be up to 6 m deep, with a correspondingly reduced plan area. Biotower filters are typically used at high hydraulic and organic loading rates to achieve a 40-70% BOD reduction.

Filter Structure

Stone or similar mineral medium filters are usually not more than 2 m deep and are contained within reinforced concrete walled enclosures. The floor is graded to provide under-drainage to a central or peripheral collector channel.

Biotower filters, incorporating self-supporting plastic media, are generally protected by a sheeting-clad steel framework which absorbs wind and other forces.

Ventilation

The ventilation of filters takes place by convective air movement through the medium due to a temperature-related density difference between the filter air and the ambient air. It is therefore important to provide adequate air access openings at the base of the medium to facilitate natural ventilation. Such ventilation is conveniently provided by designing the underdrain collector system to flow partly filled, so that it also acts as an air distribution system, and by ensuring that it is adequately vented. A total vent area of 0.4% of the plan area of the filter has been proposed (Metcalf and Eddy, Inc. 1991) as a design guide.

Filter Media

The essential requirements for a filter medium are that it should be inert, be of sound mechanical strength, and possess within its bulk an extensive area of exposed surfaces over which the liquid to be treated can be passed. Adequate void spaces must exist between adjacent surfaces to allow for some accumulation of biological film, for free passage of liquid and suspended matter, and for access of air. Performance can be correlated with the specific surface of the medium, provided that the degree of film accumulation is not such as to cause clogging. Thus with low film conditions, a 25 mm medium of a given type will produce a better effluent than a 50 mm medium of the same type. Also, for a given size of medium, rough-surfaced materials such as clinker have a marginally superior performance to that of smooth materials such as gravels. In practice, a compromise between the conflicting requirements for large specific surface and adequate void space results in the use of a medium of about 50 mm size in natural aggregate or crushed rock material. Specifications for filter media can be found in BS1438: 1971 and ASCE Manual 13. A variety of lightweight plastic filter media has been developed. Some representative types are compared with stone media in Table 2.

Table 2 : Filter media characteristics.

Medium type	Size/arrangement	Specific surface (m^2 m^{-3})	Voids(%)
Blast-furnace slag	80-130 mm, random packing	40	50
Smooth rock (basalt)		40	53
Plastic 1	PVC foil modules	85	98
Plastic 2	Polystyrene sheets, close-packed	187	94
Plastic 3	separate PVC tubes, continuous vertical lengths	220	94
Plastic 4	Polystyrene sheets	82	94

Note : Plastic media 1 to 4 are representative of available proprietary systems.

The large voidage and predominantly vertical orientation of the surfaces of plastic media, which discourages the development of thick film layers, permit unrestricted flow of liquid over the medium surface. It is probable that macroinvertebrates play only a minor role in plastic media filters, which are continuously wetted at a high rate from a fixed manifold distribution system.

COMPARISON OF ACTIVATED SLUDGE AND BIOFILTER PROCESSES

The main advantages of biofilter processes over activated sludge (AS) systems are:

(1) Because it is an attached film process, the biofilter is not susceptible to the 'sludge bulking' phenomenon, which can greatly reduce the efficiency of the AS process.

(2) In its low-rate form, the aerobic biofilter uses less energy than the AS process.

The main disadvantages are:

(1) To achieve the same degree of BOD removal, aerobic biofilters require a much larger process volume than the AS process; pre-sedimentation is also required for wastewaters containing settleable solids. Hence the capital cost of biofilter systems may be significantly higher than that of AS systems.

(2) Biofilter systems are operationally less flexible than AS systems; biofilter systems suffer a greater reduction in performance at low temperature than do AS systems.

(3) Under certain conditions, biofilter systems may give rise to odour nuisance (high-rate systems) and fly nuisance (low-rate systems).

Nitrification

As noted earlier, reduced nitrogen species are converted to nitrite, which is further converted to nitrate, in low-rate biofilters. The variation in ammonia concentration with distance from the surface of a low rate biofilter treating domestic wastewater. The extent of nitrification in biofilters is very much influenced by the organic space loading rate. The degree of nitrification is seen to decrease approximately linearly with increasing organic loading rate, reducing from in excess of 90% at organic loading rates less than 0.1 kg BOD_5 m^{-3} d^{-1} to about 40% at an organic loading rate of 0.4 kg BOD_5 m^{-3} d^{-1}. It would appear that very little nitrification takes place at organic loading above 0.6 kg BOD_5 m^{-3} d^{-1}.

SLUDGE PRODUCTION IN BIOFILTERS

There is not much published data on the rate of sludge production in aerobic biofiltration processes. The rate of humus sludge production in low-rate pilot plant biofilters has been reported (Bruce and Boon, 1970) to vary within the range 0.08-0.5 kg per kg BOD_5 removed, the higher value relating to spring time, when there is a marked shedding of filter slime and the lower value relating to summer time. The corresponding mean rate of film discharge was found to be 0.22 kg per kg BOD_5 removed, which agrees reasonably well with observed activated sludge production rates in extended aeration systems. The average rate of sludge discharge from high-rate pilot plants, treating settled domestic sewage, has been found (Bruce and Boon, 1970) to be in the range 0.63-1.0 kg per kg BOD_5 removed, with little seasonal variation in this range.

ROTATING BIOLOGICAL CONTACTOR

The rotating biological contactor (RBC) is an attached biofilm system which typically consists of a series of circular discs (biodisk) mounted on a horizontal shaft placed in a semicircular trough with approximately 40-45% disk submergence. Alternatively, a cylindrical mesh drum filled with random packing (biodrum) may be used. The biodisk or biodrum is rotated at a speed that allows adequate attached biofilm development. Oxygen transfer is achieved by the exposure and renewal of air-water interfaces as the contactor rotates and the wastewater lifted out by the rotating device trickles back down into the sump. This cyclic immersion of the biofilm also provides the opportunity for the adsorption and uptake of organics from the wastewater.

Although the concept was first put forward by Weigand in 1900 and subsequently tested by Doman in 1929, the process was not commercially developed as a wastewater treatment process until the 1960s (Huang and Bates, 1980).

While many types of proprietary RBC systems have been developed, mainly for small-scale application (< 500 PE), they generally conform to the following design parameter ranges:

disc or drum diameter:	0.7 – 3.5 m
rotational speed:	0.5 – 10 rpm
biofilm loading:	3.0 – 6.0g$BOD_5m^{-2}d^{-1}$

As in conventional biofiltration processes, the RBC process requires pretreatment by primary sedimentation and post-treatment by secondary sedimentation. Experimental biodisk studies (Bruce and Merkens, 1975), treating settled domestic sewage, have shown that where the biofilm loading is less than about 6 g BOD_5 m^2 d^{-1}, a settled effluent BOD_5 less than 20 mgl^{-1} can be obtained. The trough volume is determined by the design biofilm loading rate and the strength of the settled wastewater. It has been found advantageous (Pike *et al.*, 1982) to subdivide the trough into a set of chambers in series, thus simulating a plug flow regime and hence increasing the average conversion rate for the process unit.

RBCs are vulnerable to freeze-up and subsequent mechanical damage in cold weather and hence are normally covered in climates subject to freezing conditions.

ANAEROBIC BACTERIAL GROWTH

The Biochemistry and Microbiology of Methanogenesis

Anaerobic fermentation is brought about by populations of complex bacteria which, in very specific environmental conditions (redox potential of about -250 mV, and an almost neutral pH), form stable associations. The process is strictly anaerobic and results in the formation of methane.

This process takes place in the natural environment when a high concentration of OM occurs in the anaerobic state: marshlands, lake sediments, digestive tracts, etc.

In a simple example, such as glucose, the general equation for anaerobic digestion may be written as follows:

$$C_6H_{12}O_6 + 0.2NH_3 \rightarrow C_5H_7NO_2 + 2.5CH_4 + 2.5CO_2 + 0.6H_2O$$

The methane fermentation of OM takes place in ecosystems that are "cold" (10-15°C), mesophilic (30-40°C), and even thermophilic (> 45°C).

The various forms of degradation of complex organic matter in the anaerobic state may be described in the following manner :

- **Hydrolysis and acidogenesis phase** : This phase is performed by a wide variety of species: mesophilic, thermophilic, obligate or facultative anaerobes.

  This first phase results in a mixture of volatile fatty acids, such as acetic, lactic, propionic, butyric, etc., neutral compounds such as ethanol, gaseous products such as CO_2 and H_2, and ammonium.

  These microorganisms often have shorter generation times than those in the following phases.

- **Acetogenesis phase** : This singular acetogenesis (acetate production) phase is brought about by bacteria that are obligate producers of hydrogen (or obligate proton reducers). The process uses reduced merabolkes from the hydrolysis and fermentation phases (lactate, ethanol, propionate, butyrate). The dehydrogenation of these compounds is a process which, in the absence of microorganisms capable of using hydrogen (or a combination of hydrogen with sulphur, for example, to form H_2S), is thermodynamically unfavourable, if not impossible:

$$\text{Lactate} + H_2O \rightleftarrows \text{acetate} + 2H_2 + CO_2 + 4.18\text{kJ.mole}^{-1}$$

$$\text{Ethanol} + H_2O \rightleftarrows \text{acetate} + 2H_2 - 9.6\text{ kJ. mole}^{-1}$$

$$\text{Butyrate} + 2H_2O \rightleftarrows \text{acetate} + 3H_2 + CO_2 - 76\text{kJ.mole}^{-1}$$

However, at very low partial pressures of hydrogen, the reaction becomes thermodynamically possible and its energy variation is sufficient to allow for the synthesis of ATP and bacterial growth.

Thus, this phase is sensitive to the presence of hydrogen. It follows that to break down propionate, which is thermodynamically the most unfavourable agent, the reaction is only possible at a partial hydrogen pressure of less than 10 Pa.

- **Methanogenesis phase (in the strict sense) :** Two general methods of methanogenesis are identifiable. In the first, the H_2/CO_2 couple forms H_2O and CH_4. In the second, which is known as acetate-cleaving, acetate is broken down into CO_2 and CH_4. The second method produces about 70% methane.

  Other sources of carbon such as methanol, formaldehyde, and methylamines may also be used by methane-producing microorganisms.

THE NATURE OF METHANE-PRODUCING BACTERIA

Methane-producing bacteria are obligate anaerobes. They are characterized by the presence of co-enzymes or very specific factors such as factor F 420, whose fluorescent properties allow these bacteria to be seen under a microscope equipped with UV. Factor F 430 contains nickel, an element indispensable to the growth of these populations are similar to those in trickling filters, except that they are submerged. The high void ratio of these materials substantially limits the risk of clogging, while the relatively high developed surface area allows operation at satisfactory loadings, generally greater than 10 kg/m^3.d of COD.

- **The use of moving support materials :** Moving supports usually have a particle size of a few hundred microns. They are expanded by an ascending flow of water. A very good substrate-culture exchange and a high concentration of active biomass can be achieved with these reactors. The loadings here may reach 50 kg/m^3.d of COD or even more for some effluents.

  It is necessary to control the growth of the bacterial mass so as to avoid an excessive buoyancy increase of the support material or congestion of the bed.

  Methane fermentation is well suited to certain fluidized or expanded bed reactors, but the hydraulic system must be perfectly controlled.

LARGE SURFACE AREA PROCESSES

Large surface area processes involve purification processes with low concentration of purifying organisms in the biological reactor. These systems do not include recirculation of bacterial liquor or a separate clarification stage. They require a substantial surface area, but installation of the treatment unit is always very simple and requires little electromechanical equipment.

VARIOUS TYPES OF LAGOONING

Natural Lagooning (Aerobic)

Rays of sunlight are the source of energy enabling aquatic food chains (trophic chains) to produce living matter.

Aerobic bacteria, which are oxygenated by the photosynthetic action of vegetables that also take part in the direct synthesis of organic matter, are mainly responsible for the purification of effluents.

Aerobic bacteria include :

- microphytes, or microscopic algae, which are mainly green or blue algae that are difficult to separate;
- macrophytes, or macroscopic plants, which include free forms such as duckweed and fixed forms such as reeds.

Water hyacinths may take root. Superior vegetables act as support and normally promote the proliferation of purifying bacteria and algae.

If macrophytes attached to rhizomes develop after planting, they will exchange nutrients with the soil, the sediment, and the water. A host of predatory fauna of bacteria, phytoplankton, etc. proliferates inside the lagoons: Protozoa, Cladocera, Copepoda. Fish life may be abundant in the areas downstream of the lagoons. A natural lagoon is a very complex biological structure, where balance is not always easily achieved.

Aerated Lagooning

Oxygen is supplied artificially by floating or fixed mechanical aerators or by air injection. Unless it is adequately compartmentalized, the biological reactor is similar to a completely mixed system. The biological equilibria resemble those in the traditional activated sludge process although some algae growth is inevitable.

The microorganism concentration is low and the settling of bacterial floe is poor. Aerated lagoons are often followed by huge settling lagoons, some of which are equipped with facilities to remove the sludge. The mixing action of aerators in the aerated zones (2-5W per m^3 of lagoon) greatly reduces the formation of deposits as compared with natural lagoons.

Anaerobic Lagooning

In these lagoons, the expected purification efficiency depends mainly on the development of methane fermentation. For this reason, this process is only applied to very concentrated effluents and usually as a form of pre-treatment before a second aerobic purification stage.

The risk of pollution is particularly high and these lagoons become ineffective at low temperatures. Therefore, they may be used only in sufficiently remote areas and under favourable climatic conditions. Measures taken to cover the lagoons providing the means of recovering the biogas and improving the aesthetics of the system, have been applied to certain industrial effluents.

In practice, classifying various types of lagoons as either aerobic or anaerobic is somewhat artificial. In particular, the liquid medium of aerobic lagoons is not oxygenated in its entirety.

In the zones upstream of an aerobic lagoon, especially in the case of a natural lagoon that is compartmentalized, the pollution brought by the inflow of raw wastewater produces an oxygen deficiency.

Moreover, in the deeper layers, deposits made of the heavier suspended solids and of residue from biological activity gradually collect. Within the deposits, anaerobic fermentation phenomena occur, organic substances are mineralized and CO_2, CH_4, and possibly H_2S, are released. These deposits are particularly significant in the absence of mixing and where the wastewaters contain a great amount of easily settleable suspended solids.

In any lagoon, the aerobic and anaerobic phenomena are more or less related. If, in an aerobic lagoon, a large pare of the liquid mass is deprived of oxygen and is the scene of anaerobic bacterial growth, the lagoon is referred to as facultative. In the upper water layers, organic matter is removed through the action of aerobic bacteria and algae, while in the layers lying below the surface, it is removed by anaerobic bacteria and so-called "facultative" bacteria, which are bacteria that are adapted to both environments.

PERFORMANCE AND DESIGN

The purpose of lagoons is to remove organic pollution, reduce bacteriological pollution and sometimes to nitrify treated effluent. Temperature is an essential parameter, particularly with regard to bacterial decontamination. There is no simple mathematical model capable of explaining the purification kinetics of the complex biological phenomena occurring in natural lagoons.

Natural Lagoons

Natural lagoons are sufficiently shallow to enable an adequate amount of light to penetrate. Microphyte lagoons should be no deeper than 1.2 to 1.5 m. On the other hand, they must be deeper than 0.80 m to avoid the proliferation of macrophytes. If, on the other hand, the desired intent is to facilitate the growth of these macrophytes, as in the case of a polishing lagoon for example, a depth of no more than 0.30m is recommended.

In the temperate climate of France, the lagoon size is based on a daily loading of 50 kg of BOD_5/hectare.day, which corresponds roughly to 10 m^2 per capita and a retention time of 50 to 60 days. This basis is commonly used to convert about 90% of the BOD_5 at a temperature of 10-15°C. The level of suspended solids in the created effluent remains high and varies considerably (from 50 to 150 $mg.l^{-1}$) depending on the season. This figure rises steeply with increased insolation. The total BOD_5 of a municipal effluent treated by lagooning is therefore usually greater than 50 $mg.l^{-1}$.

In polishing lagoons situated downstream of a conventional activated sludge treatment facility and designed to reduce bacterial contamination (a reduction of 3 to 4 logarithmic units of bacterial concentration expressed in indicator germs), a retention time of about 30 days is usually adequate under typical conditions in France as long as the temperature does not drop below 15-20°C. This temperature is also very important if nitrification is required.

Aerated Lagoons

Aerated lagoons may be as deep as 2.5 to 3 m. The maximum acceptable depth depends on the stirring power of the aeration system.

One approach to the necessary retention time is based on the application of Monod's model for a single completely mixed reactor. If So and Sf are BOD_5 levels in the raw and treated wastewaters, then:

$$\frac{Sf}{S_o} = \frac{1}{1 + K_{Tt}}$$

where,

t is the retention time in days,

K_T depends on the temperature T in °C

according to the relation $K_T = K_{20} x$

1.07^{T-20} if T > 10°C. For municipal wastewater, K_{20} is about $0.5.d^{-1}$.

The temperature T_1 in the lagoon may be calculated by the equation:

$$(T_o - T_1) = \frac{(T_1 - T_a) FA}{Q}$$

where,

T_o is the temperature in °C of the wastewater to be treated,

T_1 is the temperature in °C in the lagoon,

T_a is the temperature in °C of the ambient air,

A is the surface area in m^2 of the lagoon,

Q is the flow in $m^3.h^{-1}$,

F is a coefficient that varies from 0.5 to 1.2 in the temperate zones of the northern hemisphere.

In reality, dividing the lagoon into compartments and rational distribution of aerators reaps some of the benefits of having the ranks arranged in series. On the other hand, the dispersed character of the bacterial floc in low concentrations (low sludge age and limited level of stabilization) and the development of algae are detrimental to good separation of suspended solids. In the treatment of municipal wastewater, the SS concentration in the treated effluent may vary between 50 and 250 $mg.l^{-1}$.

Locating a seeding lagoon downstream of the aerating lagoon allows these values to be reduced. The soluble BOD_5 of the treated effluent is low and corresponds to a reduction of 80 to 90% for a retention of about 10 days in temperate climatic conditions. The efficiency of bacterial decontamination depends mainly on the retention time and the absence of preferential paths in the lagoon.

It is therefore clear that aerated lagoons are best suited to the breakdown of organic pollution and that natural lagoons are particularly well suited to reducing the bacterial concentration in an effluent char has already undergone secondary purification, whence the name 'stabilizing' or 'ripening' lagoons.

Anaerobic Lagoons

The retention times are greater than 20 days and often exceed 50. The BOD loadings are about 0.01 $kg/m^3.d$ and the removal efficiency may vary in a wide range of 50 to 80%. The SS level is high. Deep lagoons (5 to 6 m for example) should theoretically enhance the process.

Design and Maintenance of Lagoons

The compartmentalization and operation of lagoons in series tends to improve efficiency. This layout allows for successive facilities corresponding to different ecosystems and/or purifying procedures. However, extensive anaerobic phenomena and foul odours may occur if the raw effluent is concentrated.

A geotechnical study is indispensable before selecting a lagoon process solution. Also the reservoir must be watertight. The banks must be as steep as possible and be protected against wave action (adequate rock support, sufficient freeboard) especially in aerated lagoons.

The operation of the lagoons is made easier if settleable and floating matter is previously removed from the wastewater. The ideal solution for municipal wastewater consists in installing upstream of the lagoon a pretreatment and primary seeding facility combined with, for example, an anaerobic digestion plant. However, economic factors generally dictate that the lagoon be supplied with wastewater that has not been previously settled. The upstream zones of lagoons often carry large amounts of sludge deposits that may cause a nuisance.

The systematic cleaning of lagoons is an essential part of operations and it is important that this be taken into account in the study phase. Sludge removal is often accompanied by the clearance of excess aquatic plants, particularly in natural lagoons.

The use of Soil for Purification

The first method used to purify wastewater was land disposal. Using soil for purification takes advantage of:

- the physical and physical/chemical properties of the soil, including filtration, adsorption, ion exchange, retention capacity;
- the biological properties of the soil, including the action of the microflora and planes.

The purification system is thus made up of both soil and plants. Some of the polluting elements are released into the atmosphere. Some of the carbon is transformed into carbon dioxide by bacterial respiration and by photosynthesis. Other polluting elements are removed by plants. Firstly, C and N, then, P, K, Ca, Mg and finally, to a lower extent, heavy metals.

The three main treatment methods using soil are irrigation, infiltration-percolation and controlled runoff.

Table 3 gives typical characteristics and performance of these treatment systems.

Irrigation is the most widespread system. Water is supplied through ditches or sprayed. The underground water table must not be located at a depth of less than about 1m. The soil must be moderately permeable and drainage is frequently recommended. It is clear that the quantities of water employed vary according to the type of plane growth, the nature of the soil, and the climatic conditions. The amount of water applied, including precipitation, muse correspond to the natural evaporation rate and plane requirements (evapotranspiration and growth). This figure varies throughout the year. The balance of nutrients must also be taken into account. The organic loads usually applied range between 5 and 20 kg of BOD_5 per hectare per day.

Apart from the occupation of extensive areas of land, the main difficulty in applying these irrigation techniques is the limited period of use. For annual crops, land disposal is only possible during 3 to 6 months per year. In grassland, land disposal is not possible during the rainy season or in the months before animals are put out to pasture. Tree plantation (poplars) are more tolerant. A detailed plan of irrigated crop rotation has to be drawn up.

It is preferable that the raw wastewater has at least undergone settling before use. The more advanced the pretreatment of the effluent, the more constant and rapid its infiltration rate.

Table 3 : Characteristics and performance of the main methods of treating municipal wastewater by soil.

Treatment	Annual hydraulic loading m/year	Daily hydraulic loading mm/day	Required pretreatment	Treated water mg.l^{-1}		
				BOD_5	TKN	SS
Irrigation	0.5 to 3	3 to 10	Primary settling	< 5	< 5	3 – 5
Infiltration-percolation	4 to 50	20 to 200	Primary settling	< 5	< 5	10
Controlled runoff	1 to 10	10 to 30	Screening	< 20	< 15	5

SEWAGE TREATMENT

Treatment Parameters

A typical chemical analysis for raw sewage is shown in Table 4. For the UK the Royal Commission Standard of 30mg/l solids and 20mg/l BOD has been the established target for discharge to watercourses. However this standard was based on a minimum dilution by the receiving waters of 8 : 1. Increasingly with ever higher abstraction and water usage many watercourses cannot provide this dilution. In addition, especially in arid countries where water is scarce, re-use of the treated effluent is required. Both situations demand more stringent levels of treatment. Thus, primary and secondary stages are being augmented by tertiary treatment to achieve levels of 10mg/l solids and 10mg/l BOD or better. The most common laboratory test used for sewage is the BODs test (the five-day biochemical oxygen demand at 21°C). This test gives a measure of the concentration of biodegradable organic material in the sewage or effluent and is effected by measurements of the dissolved oxygen removed from the water by the microorganisms while feeding on the organics. A sewage sample is diluted with oxygenated water and the dissolved oxygen concentration measured. The sample is placed in an incubator (temperature-controlled container) at 21°C for 5 days. At the end of this period the oxygen concentration is measured again and subtracted .from the original reading to give the amount consumed by the microorganisms. This is then adjusted by the dilution factor to give the BOD_5.

Table 4 : Average (24h composite) analysis of crude sewage.

Analysis	Concentration (mg/l)
COD	650
BOD	326
Organic carbon	173
SS	127
Organic N	19
Ammoniacal N plus urea nitrogen	47
Total nitrogen	47
pH (value)	66
Anionic detergents	7.8
Sodium	100
Potassium	20
Calcium	20
Magnesium	110
Chloride	70

Ammonia (NH_3) is also present in the sewage as a result of natural breakdown of urea (urine). Ammonia is not organic because it does not contain carbon. It is not reduced significantly by the same microorganisms that remove organic materials. It is removed mainly by nitrifying microorganisms (*e.g.* nitrobacter). This process is called nitrification and results in the conversion of ammonia to nitrate.

It is important when ammonia removal is required that for nitrification to be complete it is necessary for the sewage to contain at least 8mg/l alkalinity (as $CaCO_3$) for every mg/l ammonia removed (as N). This can be a problem with sewage derived from desalinated waters. In such designs plant modification may be necessary (addition of alkalinity and/or denitrification).

Another common laboratory test used for sewage is the suspended solids test (SS). This test measures the concentration of undissolved solids present and is effected by filtering the sample and weighing the solids deposited on the filter paper.

Daily Variations in Sewage Flow

Fortunately for the designers of sewage-treatment plants, sewage characteristics are fairly consistent world-wide. Unfortunately the pattern of everyday life introduces large variations in sewage flow over the 24 hours of a day. Older combined foul and surface water collection systems will show peak levels during high rainfall. A community comprising mainly facilities, with husbands going to work, children going to school and most wives staying at home, does have a fairly predictable sewage-flow pattern.

In this case most of the sewage will be discharged during 16 hours per day with very little flow during the 8 hours during the nights. The flow will peak during the morning, after lunch-time and in the early evening. At these three peak periods the flow could typically be three times the average daily flow, each time for about one hour.

In contrast to this typical domestic community is, for example, the all-male workers camp. In an extreme case all of the daily sewage could arrive during the two-hour periods which are early morning and early evening at a flow of perhaps 6 times the average daily flow.

It follows that the larger the community served by a treatment plant the greater will be the attenuation of localized peak flows and the ratio of peak average flow to be treated will be reduced.

Thus, without collection and flow balancing, all treatment plant carries the burden of catering for wide flow variations in its hydraulic engineering. Fortunately the active biomass utilized in most treatment plants has the ability to absorb pollutants rapidly on contact and the controlling parameter is not hydraulic residence time. Typical values for sewage flows are given in Table 5.

Sewage Treatment Processes

There are many processes that can be used to purify sewage:

- *physical processes* include screening, settlement, flotation, sand filtration, membrane filtration, reverse osmosis and distillation;
- *chemical processes* include coagulation and flocculation using polyelectrolytes, lime, iron salts (ferric chloride or ferrous sulphate) or aluminium salts (aluminium sulphate), oxidation using chlorine, hypochlorites, ozone or peroxides and adsorption on activated carbon;
- *biological processes*
  - where microorganisms are freely suspended in the water, such as activated-sludge processes including extended aeration and contact stabilization
  - where microorganisms are attached to a surface, such as biological (trickling, percolating) filters or filter beds, and rotating biological contactors (biodisc, biosurf, biospiral etc.).

Table 5 : Guidelines for the average flow of sewage from various establishments.

Type of establishment	Volume of sewage (l/head/day)
Small domestic housing	120
Luxury domestic housing	200
Hotels with private baths	150
Restaurant (toilet and kitchen wastes per customer)	30-40
Camping site with central bathhouse	80-120
Camping site with limited facilities	50-80
Day schools with meals service	50-60
Boarding schools – term time	150-200
Offices - day work	40-60
Factories – per 8 hour shift	40-80

All of these processes have been used at some time, either in a full-scale plant or on an experimental basis and Figure 5 shows the layout of a large treatment plant. From the results of plant operation, experiments and the associated capital costs and running costs certain economical combinations of processes have become preferred for the purification of domestic sewage:

Preliminary treatment : Simple physical processes at the plant inlet to remove materials that may damage, block or inhibit the main treatment processes. These are usually screening (to remove debris) or comminution (to shred debris), often sand and grit removal by flotation and sometimes flow equalization.

Primary treatment : Removal of settleable solids from screened sewage usually by settlement but occasionally by very fine screening or flotation. Primary treatment produces an odorous and infectious sludge (primary sludge) which usually requires further treatment. For this reason it is often omitted and the downstream processes modified to suit the additional solids load (as in extended aeration).

Secondary treatment : Removal of organic materials from screened or settled sewage by biological processes. Aerobic microorganisms are much more effective than anaerobic in the proportion and speed of BOD removal. Also, aerobic conditions assist in the removal of pathogenic microorganisms. Aerobic processes require oxygen to be dissolved in the water. This can be achieved by the injection of pure oxygen but it is more common and economic to use mechanical (surface) aerators or diffused (bubble) aeration to transfer oxygen from the air.

Biological processes convert biodegradable organic materials (BOD) to more microorganisms and carbon dioxide (a gas which escapes to atmosphere during aeration). Most of the microorganisms in the effluent are removed by settlement (secondary settlement) or sometimes by very fine screening. All biological secondary-treatment processes produce more micro-organisms which require disposal, sometimes after further treatment (Digestion and dewatering). The excess microorganisms are referred to as secondary sludge (excess, surplus or waste-activated sludge or humus sludge).

Tertiary treatments are the traditional reference for all further processes following secondary treatment. These processes often are those also used for water treatment. The tertiary treatment required, if any, is dependent on the final destination or re-use of the treated effluent, or needed to meet a required analysis (BOD, suspended solids, bacteria etc.). This is usually achieved with sand filtration (gravity type or pressure type) sometimes assisted with anthracite. Most are disinfected before disposal and this usually means dosing with chlorine or hypochlorite (sodium or calcium). Sometimes, tertiary treatment is attempted with microscreens (drum screens, fine screening).

A typical secondary effluent of 20mg/l BOD and 30mg/l SS given treatment by sand filtration should produce a filtered effluent of 10mg/l BOD, 10mg/l SS. To improve this quality to 5mg/l BOD, 5mg/l SS further treatment through granular-activated carbon filters would be commonly chosen.

To achieve higher standards than this, chemical dosing and clarification (settlement) using typically polyelectrolytes, alum or ferric chloride, need to be interposed between secondary treatment and sand filtration. A good system could produce 1mg/l BOD, 1mg/l SS.

To produce water of drinking quality with zero concentration of BOD, SS and coliforms (bacteria) it is necessary to interpose ozone treatment between secondary treatment and chemical clarification, and include RO after sand filtration. Activated carbon would not be used as its function is replaced by ozone and RO, and it sometimes generates biological solids which may foul the RO membranes.

NATURE OF SUSPENSIONS IN SEWAGE TREATMENT

The suspensions encountered in sewage treatment include the raw wastewater itself, the humus and activated sludge suspensions produced in biological treatment processes, and the flocculent suspensions generated by physicochemical treatment.

Raw municipal sewage, which is a mixture of domestic and industrial wastewaters, varies in composition from hour to hour, day to day, and location to location. The total solids content of raw sewage (*i.e.* the sum of the dissolved, colloidal and suspended species of solid matter) depends on the per capita water consumption and the nature of the industrial discharges in the sewer catchment area. It varies in the typical range 500-1000 mgl^{-1}. A typical fractional characterization of the solid matter in sewage as organic, inorganic, dissolved and suspended species.

The inorganic particulate fraction of municipal wastewater, commonly referred to as grit, which typically has a much higher specific gravity and settling velocity than the organic particulate fraction, is conventionally selectively separated in the first stage of treatment. The remaining particulate matter varies in size from 10 mm to colloidal size. Excluding the grit fraction, the specific gravity of the dry solids is in the range 1.0-1.5 but wet solids contain a high proportion of water bringing their specific gravities close to 1.0. Roughly 55-65% of the total suspended solids in municipal wastewaters are settleable and hence removable by a gravity-settling process commonly called primary sedimentation. The nature and volumetric concentration of settleable solids in sewage are such that their settling in primary sedimentation processes may be categorized as hindered flocculent settling. Primary sedimentation generates a solids residue or *primary sludge* which can be readily thickened to achieve a total dry solids concentration in the range 4-8% by weight.

Activated sludge is the microbial suspension or so-called 'mixed liquor' generated in aerobic suspended floc biological treatment processes. It is more consistent in quality and concentration than crude sewage but varies somewhat with operating conditions. The activated sludge particles or flocs range in size from a few microns to more than 1 mm. The specific gravity of the dry solids is over 1.5 but, since this is an organic flocculent suspension containing much entrained water, the in situ specific gravity ranges from 1.01 to 1.03. The mixed liquor suspended solids concentration may vary in the range 1500-5000 mgl^{-1}; being typically highly flocculent, it exhibits the zone-settling behaviour. However, some activated sludges flocculate poorly and hence have very low settling velocities. This condition is termed 'bulking'.

Humus sludge is the microbial biofilm residue generated in aerobic attached film biological processes (aerobic biofiltration). The suspension varies in particle size from colloidal to macroscopic with considerable fluctuation in nature and concentration which are subject to seasonal change. The typical concentration range is 50-150 mgl^{-1} with a specific gravity similar to the organic fraction of crude sewage. The percentage of suspended solids which are settleable is usually in the region of 70-80%.

All these suspensions include colloidal particles which are not amenable to separation by the conventional separation processes. The characteristics of colloidal suspensions and the processes that can be used to effect their destabilization.

NATURE OF SUSPENSIONS IN WATER TREATMENT

The colloidal impurities of natural surface waters are made up of a complex mixture of particles consisting mainly of various colloidal clays. Depending on the degree of contamination, surface waters may also contain colloids from domestic and industrial wastes, live and decaying algae and their decomposition products, bacterial cells, decaying organic matter, and colour colloids. Some of the colloidal impurities, such as clay, may be hydrophobic in nature, whereas others, like certain sewage colloids, may be hydrophilic.

The size of the particles concerned lies between the size of water molecules and the visible matter suspended in the water. Colour colloids are very small and almost approach the size of water molecules. They have been shown by Black and Christman (1963) to be mainly in the 4-10 nm range. Clay particles generally fall in the 0.1-2 pm range. Because of their small size, these colloidal particles are not visible under an ordinary microscope, even of the highest power. They may be viewed under an electron microscope or by means of the Tyndall effect.

The process of coagulation leads to destabilization and aggregation of colloidal particles into hydroxide floc particles of size up to 1 mm and above, of increased density and settling velocity. These floc particles are of loose structure and contain much entrained water. The coagulating compounds have the following dry densities :

Al oxide	Al_2O_3 (20 H_2O)	1180 kg m^{-3}
Fe oxide	Fe_2O_3 (20 H_2O)	1340 kg m^{-3}
Crystalline	$CaCO_3$	2600 kg m^{-3}

Their wet densities depend on the amount of entrained water and may calculated using equation.

Experiments reported by Sterina (1964) give some indication of the effect on the floc of the addition of a long chain polyacrylamide. Artificially turbidified water was used, containing a suspension of clay and floc densities ranging from 1019 kg m^{-3} to 1043 kg m^{-3} were reported. Similar experiments with coloured waters gave values ranging from 1001.5 kg m^{-3} to 1004.7 kg m^{-3}.

With regard to the amount of floc produced, it is reported that the ratio of floe volume to weight of coagulant cation is variable according to the natural water in the range 100-300 ml floc g^{-1} cation. Hudson (1965) found that 1 g Al^{3+} yielded 250 ml floc and 1 g Fe^{3+} produced 250 ml and this value was found independently by Ives (1968). There appears to be no standard method of determining floc volume.

In experiments on an upflow sludge blanket clarifier, Ives and Hale (1970) determined volumetric concentrations in the sludge blanket by lowering a 50 mm diameter tube into the suspension and withdrawing the tube containing the floc from a known location. The floc in the tube was then tested for settlement rate, volumetric concentration and fluidization characteristics. The volumetric concentration was calculated from the volume occupied after 3 h quiescent settlement. It was found that the volumetric concentration varied from 0.20 to 0.05, corresponding to nominal upflow rates ranging from 1 to 2.5 m h^{-1}.

Sewage Treatment

Water-borne Discharges

Treated sewage is over 99.9% water and contains a wide variety of substances in small or trace concentrations. 96% of the UK's 55 million population is connected to public sewers and 83% of this is served by sewage treatment works. The remaining 17% discharges to sea through marine outfalls, often without even preliminary treatment.

Discharges to inland waters are controlled mainly by the NRA to prevent overloading the receiving waters with organic polluting matter. This maintains dissolved oxygen levels and protects the flora and fauna which often thrive only within a narrow range of environmental conditions. In severe cases of pollution the result could be no aerobic life at all. For discharges to oxygen rich marine waters the biochemical factors are not so important and bacterial content can be the limiting criterion.

The majority of Water Resources Act consents are long standing many having first originated from the Rivers (Prevention of Pollution) Acts of 1951 and 1961. The standard requirement for most sewage effluent is for a limit on five day BOD and Suspended Solids (SS) with a lesser number of sensitive discharges requiring a limit on Ammoniacal Nitrogen (AmmN). The basic treatment requirement for most inland works, therefore, is primary settlement, a biological treatment stage, and secondary settlement. The degree of treatment necessary in the second stage will depend upon whether an ammoniacal nitrogen limit has been applied. The values of quality parameters are determined by the NRA after assessing the volume of discharge and quality objectives for the receiving watercourse, taking into account the dilution available.

The percentage of sewage works in England and Wales failing to comply with consent conditions improved from 23% in 1986 to 6% in 1991. The water industry has spent around £ 11 billion over the period 1989-93 since privatization and is currently committed to further improvements.

The formats of consents to incorporate the requirements of the UWWTD and the Water Resources Act have recently been drafted. Provision of treatment facilities under the UWWTD will be more demanding than under current law, especially for sensitive waters and coastal discharges. Standard Waste Water Treatment Directive consents for works treating a population equivalent (p.e.) of > 2000 will require reductions in BOD of 70% and Chemical Oxygen Demand (COD) of 75%, or fixed emission standards of 25 mgl^{-1} BOD and 125 mgl^{-1} COD. The optional Suspended Solids limit of 35 mgl^{-1} or reduction of 90% is not being adopted. Works treating < 2000 p.e. should have treatment appropriate for local conditions,

Discharges to 'sensitive' waters will have additional requirements for phosphorus and nitrogen, since excessive levels could stimulate plant and algal growth and alter the ecology of the water. A phosphorus limit of between 1-2 mgl^{-1}, depending upon the volume of discharge, or a reduction of 80% will be required. Total nitrogen limits of 10-15 mgl^{-1} N, or a reduction of 70-80%, may also be necessary. Discharges to esturial or coastal waters in 'high natural dispersive areas' will only require primary treatment with a reduction in BOD of at least 20% and a reduction in suspended solids of at least 50%.

To comply with the new standards, secondary treatment will have to be provided except for some small works and in areas of high natural dispersion. These areas have been agreed with the NRA, but comprehensive studies will have to be undertaken to prove that there will be no detrimental effect in not providing secondary treatment. The introduction of sensitive areas will also result in the provision of a nutrient-stripping tertiary treatment stage. The eastern side of the country is most affected since the rainfall is less and dilution in rivers is subsequently lower.

Most agglomerates of > 2000 p.e. are already served by a sewerage system. However, as most of these sewers are combined, intermittent storm discharges can adversely affect the receiving watercourses. Under the UWWTD remedial work must be undertaken to unsatisfactory overflows by provision of additional sewer capacity and/or screening to remove aesthetic pollution by plastics and other debris from the watercourses.

The provision of sewage treatment works for esturial and coastal populations, which currently may have no treatment, will mean high capital expenditure and increased operational costs: Capital expenditure and increased operational costs will also result from the provision of nutrient removal and storm overflow improvements.

The EC Dangerous Substances Directive and the related regulations, will impose much tighter control on prescribed substances discharged in sewage treatment works effluents, to prevent toxic substances from entering the aquatic environment. Strict trade effluent control at industrial premises is the only economic approach to this problem.

Sludge

Sewage sludge is the mainly organic residue remaining after sewage treatment. Sewage originates from a wide diversity of households and industrial activity and so sludge constituents can vary considerably over a short timescale.

The sludge is usually treated before disposal by a combination of the following processes :

(i) gravity or mechanical thickening to reduce the volume of the liquid sludges;

(ii) anaerobic digestion to reduce the overall weight of sludge solids and reduce smell and pathogen levels;

(iii) mechanical dewatering to produce a stable solids product; and

(iv) drying by heat treatment to produce a granular or pelleted product, which is odourless and pasteurized to control bacterial levels.

Disposal methods must be economical and minimize adverse environmental consequences. Where possible, the beneficial characteristics of sludge can make recycling a practical alternative. Table compares some of the qualities of sewage sludge.

Sludge Disposal Routes

Disposal routes are strictly controlled by EC legislation and Regulations to maximize the benefits of sludge and prevent environmental damage. Table shows the quantities of sludge and disposal routes in 1991.

The Government agreed at the Third North Sea Conference in 1990 that sea disposal would cease after 1998; this is also a condition of the UWWTD. The additional sewage treatment arising from the UWWTD will result in a significant increase in sludge production, and the predicted quantities of sludge and disposal routes are also shown in Table 6. An EC Directive (86/278/EEC) aims to maximize the benefits of sewage sludge within the community whilst ensuring the potential adverse effects are carefully monitored and controlled. Some of the benefits of alternative disposal routes are discussed below.

Agriculture

This is the most common option. Liquid digested sludges are usually spread onto the surface of grasslands. Care has to be taken to prevent run-off from polluting beneficial.

(i) The water content varies from 99% down to 5%, dependent upon the amount of thickening or dewatering which takes place, and can be of benefit to a growing crop.

(ii) Provision of some essential trace elements for crop growth.

(iii) Nitrogen and phosphorus, at levels between 3% and 6% of the dry matter present, are valuable crop nutrients.

(iv) Organic matter, which varies between 50% and 80% of the dry matter present, improves the soil structure and its ability to retain moisture.

Non-beneficial

(i) Pathogen levels are high in untreated sludges but can be substantially reduced, dependent on the type of sludge treatment.

(ii) Untreated sludges smell badly and can be the cause of many complaints; dependent upon the treatment process selected, the smell problem can be reduced or even eliminated.

(iii) High levels of fats and greases cause sludges spread onto pastures to cling to the sward rather than being easily washed down to the root systems.

(iv) Large amounts of paper and plastics are unsightly and if spread to pastures may be ingested by animals.

(v) Contaminant levels, *e.g.* high metal content, may preclude safe disposal to farmland.

Untreated liquid sludges are usually either surface applied to arable land, which is then ploughed, or they are injected beneath the surface. Incorporation into the soil prevents the spread of pathogens and keeps smell problems to a minimum.

Nitrogen release is much slower, allowing it to be taken up by an arable crop. 'Cake' sludges, which are often untreated, are generally ploughed into arable land at a rate of approximately 50 tonnes ha^{-1} $annum^{-1}$.

Table 6 : Sludge disposal.

Disposal tds	1991 tds*	Estimated 1999
Agriculture	465000	777000
Dedicated	25000	30000
Sea Disposal	334000	0
Incineration	77000	382000
Landfill	88000	35000
Beneficial	68000	137000
Within Curtilage	50000	91000
Uncertain	–	294000
Total	1107000	1746000

*tds = tonnes dry solid

Within England and Wales the Directive is enacted under Regulation and further enhanced with a Code of Practice. These regulations require sludge producers usually Water Services Companies, to analyse receiving soils to a depth of 25 cm before sludge is applied to ensure that the following conditions are satisfied:

(i) the pH is > 5.0, thus minimizing heavy metal transfer from soil to plants;

(ii) the concentrations of each of seven heavy metals (Zn, Cu, Ni, Cr, Pb, Cd, Hg) are less than prescribed limits.

They also require analysis of sludge quality to ensure that :

(i) this metal loading of the seven heavy metals must not exceed prescribed limits;

(ii) the concentrations of lead and fluorine in sludges destined for grassland application must not be higher than prescribed levels; and

(iii) the concentrations and quantities of nitrogen and phosphorus are established.

This information must be recorded in a register for audit and inspection and is used to demonstrate appropriate environmental control. Information about soils and sludge application must be made available to the land owner.

The associated Code of Practice further requires that :

(i) the soil is again sampled, at a depth of either 15 or 7.5cm for pasture or arable purposes, respectively;

(ii) the concentrations of each of the seven heavy metals in the sample are less than prescribed limits;

(iii) only treated sludges may be placed on grassland and animals must not be allowed to graze there for three weeks after sludge application in order to reduce the risk of infection; and

(iv) 'cake' sludges may not be used on grassland for fear of ingestion by animals.

In order to minimize the impact of excess nitrogen on the environment and prevent high levels in water, another Code of Practice seeks to:

(i) limit the amount of organic nitrogen applied to land in any one year to 250 kg ha^{-1};

(ii) ensure that it is applied only when there will be no fear of run-off to watercourses; and

(iii) ensure that the nutrients are applied when the growing crop can use them.

Further constraints on sludge recycling to farmland include the following:

(i) the requirement to apply no sludge to land constrained by the 'set aside' scheme of the Ministry of Agriculture, Fisheries and Foods (MAFF); and

(ii) the need to add little or no sludge to land designated as 'Environmentally Sensitive'. This scheme compensates farmers who limit the amount of nutrients added and thus preserve, rather than enhance, the fertility of designated land.

In addition to agricultural use, some sludges are recycled as top dressings for tips, road embankments, and for recreational areas, such as golf courses.

Landfill

Normally landfill sites are used when farmland lies outside an economic radius for sludge recycling or when the sludge is contaminated and poses a pollution risk to the soil. Many 'cake' sludges are disposed to landfill since filter pressing significantly reduces both the volume to be transported and the reception fee charged at the landfill. The sewage treatment preliminary processes and sludge screening produces rags, plastics, wood, and grit. These materials are usually also sent to landfill sites for disposal.

The Environmental Protection Act 1990 requires that Landfill sites are registered by county based Waste Regulatory Authorities. These Authorities set strict limits and controls on the quantity and type of waste materials which can be accepted into registered sites. These limits are specific to a particular site and are designed to ensure that the materials are compatible and will be retained and treated within the site without environmental pollution.

Forestry

Sludge application in forestry is difficult. Land is often of poor quality and far from the towns which produce the sludge; it may be steep and difficult to incorporate sensible amounts of the sludge without causing run-off. Access may be restricted by inadequate road systems or even by heavy tree growth itself before the saplings are thinned.

Disposal to forestry is now covered by a Code of Practice. This code recognizes the tree growth benefits of adding nutrient and organic matter to land often of poor quality. It also seeks to ensure forest land is subject to many of the precautionary constraints which apply to sludge recycling on farmland; thus nutrient and metal loadings are controlled in a similar fashion.

Sea Disposal

Until 1998 sea disposal will remain under the careful control of MAFF, under the Food and Environmental Protection Act 1985. Disposal sites are chosen carefully and licensed by the Ministry only after passing an exhaustive checklist designed to ensure that no harm will result to the marine environment. The sludge producer must discharge the sludge at specific states of the tide and install equipment which continuously monitors the whereabouts of the vessel, indicating whether it is travelling full or empty and whether it is carrying sludge or sea water ballast. The licence limits the sludge volumes and the maximum loads of heavy metals as well as the concentrations and loads of PCBs and some organochlorine pesticides.

Incineration

The closing of the sea disposal route has forced Water Service Companies to develop new disposal strategies. Farmland disposal may not be an option for coastal cities producing large quantities of sludges, perhaps contaminated by industry. Incineration requires sludge to be dewatered to at least 30% dry matter. HMIP strictly controls the emission levels of metals, particulates, and various gases such as sulfur dioxide and carbon monoxide, and by-products of combustion, *e.g.* dioxins, under the Environmental

Protection Act 1990. Fuel gas scrubbing can minimize atmospheric emissions. Some 30% of the sludge dry matter remains after combustion as ash and this normally is removed to landfill.

Atmospheric Emissions

Sewage treatment can be an odourous process under certain anaerobic conditions. The sewerage systems and treatment works are usually designed to minimize the production of odours, which are usually sulfur based. Hydrogen sulfide odours from sewerage systems and primary processes occur in septic conditions caused by long retention periods, hot weather, and very strong sewages. Hydrogen sulfide and mercaptans can be released during desludging of primary tanks or from areas of turbulent flow if the sewage has turned septic. Hydrogen sulfide is a toxic gas and in concentrations above 10 p.p.m. can be lethal to workers in sewers and pump sumps. It is also corrosive to concrete and equipment. Treatment units and problem areas can be covered and the odorous gases treated either chemically or biologically.

A beneficial gas, methane, is produced from the anaerobic digestion of sludges. Power generation is generally considered economic for population equivalents > 100000 and sometimes for plant of > 50000 p.e. The electricity can be sold to the Electricity Industry at a premium rate under non-fossil fuel agreements. Excess methane, which is a greenhouse gas, is either flared off or released to the atmosphere. However, < 3% of the estimated methane emissions originate from sewage treatment.

Public Health

Micro-organisms are a ubiquitous and intimate part of our daily existence. Every day large numbers of them are discharged into the wastewater system from healthy and unhealthy people through diverse human activities. Risks arising from the wastewater generated will depend on the general state of health of the population. Inadequate hygiene and sanitation systems can cause contamination of foods and water used for drinking. Epidemics of cholera and typhoid can occur and spread rapidly, particularly in poor and deprived areas. The South American pandemic of 1991 and Rwanda of 1994 are recent illustrations. Hostilities such as the Gulf War can often result in the breakdown of sanitary systems and increase risk of water-borne disease.

Table 7 : Quality of bathing water 1989-92, % compliance with CEC directive.

Sea Water	1989	1990	1991	7992
Belgium	94	90	85	90
Denmark	–	–	93	95
Germany	–	–	64	73
Greece	–	–	90	97
Spain	80	–	88	93
France	84	–	87	79
Ireland	96	–	97	94
Italy	84	–	90	92
Luxembourg	–	–	–	–
Netherlands	–	–	86	86
Portugal	–	87	83	83
UK	76	77	76	79

Wastewater treatment has generally been directed towards the reduction of aesthetic and obvious polluting aspects of wastewater discharges, rather than microbiological aspects of the processes. Most treatments achieve some reductions in faecal, including pathogenic, micro-organisms and some plants may be augmented with various tertiary treatments including lagooning, sand filtration, membrane filtration, and disinfection which further reduce bacterial levels. Tertiary treatment or chemical disinfection are used particularly where microbiological standards apply in the receiving waters. These standards tend to be most depending where recreational use is to be made of the receiving water, as with inland waterways and resort bathing waters.

Microbiological standards relating to coastal and esturial waters include European directives on bathing water quality and the water quality in shellfisheries. Public and media concern about sewage disposal in these areas has related not just to compliance with these standards but also to more obvious aesthetic signs of pollution, including material washed up on beaches or floating in or on the water. This has an adverse psychological effect on the public well-being and recreational value, quite apart from any subsequent morbidity associated with faecal contamination of the water. The 'Good Beach' guide statistics and 'Blue Flag' awards include standards relating to cleanliness and provision of facilities such as toilets at the beach site, in addition to the achievement of microbiological quality standards. European compliance with the Bathing Water Directive is given in Table. Gradual improvement in the UK in bathing water quality is evident.

In fact, the microbiological standards of the Bathing Water Directive were set arbitrarily and not on the basis of knowledge of public health significance. Subsequent epidemiological studies aiming to relate standards to health criteria have been widely criticized by scientists on statistical grounds. In practice, these studies are both difficult and expensive to perform. The most recent and largest study of this kind concluded that there was a correlation between faecal micro-organism numbers (particularly Enterovirus) and gastro-intestinal symptoms. The correlation in those who had been exposed to the water was only significant, however, when counts significantly exceeded the mandatory, imperative standards of the Bathing Water Directive. Other categories of symptom such as eye, ear, nose, throat, and skin disorders were correlated with degree of water contact, duration, and intimacy of exposure, rather than concentrations of microbial indicators of faecal pollution. The Bathing Water Directive microbiological standards are being reviewed and amended, giving greater importance to faecal Streptococci in line with experience, although European consensus has yet to be achieved.

Viruses arc of topical interest in relation to exposure risks from wastewater use and discharged. Public Health Laboratory figures show that Hepatitis A, responsible for infectious jaundice, has increased since 1987. Shellfish consumption can increase risk but no link has yet been made with recreational water exposure. There is no evidence supporting risk to the public from exposure to recreational waters or sewage from HIV, the causative agent of AIDS.

Bathing waters can suffer from potential microbiological hazards other than from sewage discharge. Notably, environmental contamination can result from gulls which have a prevalence of enteric bacterial pathogens such as *Salmonella* and *Campylobacter* in their faeces.

The position is similar where inland waters are used for recreation. The concentration of faecal organisms is the principal factor in the risk of disease. Recreational exposure, including canoeing, swimming, sailing, and fishing differs only slightly from marine waters in the degree and intimacy of exposure. Rivers act as repositories for large bodies of treated wastewater, surface water, and agricultural run-off. Sewage effluents, although subject to stringent discharge consents, still have significant inputs of faecally derived microbes. Abstraction of river water for drinking purposes places constraints on the location of the water purification works and the type of treatment required.

The often quoted risks of Leptospirosis, or Weil's disease, must be kept in context. Sewage workers may have an increased risk from indirect exposure to rats and this has led to confusion with the risk of exposure to sewage. In fact the disease is usually contracted by exposure to the urine of infected rats or of urine-contaminated stagnant water. Infection occurs through cuts and abrasions or via the mucous membranes.

Environmental Impact Assessment

Sewage treatment works have an environmental impact which must be considered in the selection of the site location. This is especially true for new works on green field sites, such as many of the coastal sites required under the UWWTD.

Developments with a potential for environmental impact are covered by the EC directive 85/337/EEC concerning the effects of certain Public and Private projects on the environment. The directive was implemented in the UK through the Town and Country Planning (Assessment of Environmental Effects) Regulations in 1988 and amended in 1990 and 1994. Within the UK, planning applications for sewage treatment works are determined by County Planning Authorities under their responsibility for waste disposal or the Local Planning Authority.

Sewers and small works extensions constitute 'permitted development' under the Town and Country General Development Order and are not subject to Environmental Assessment (EA). The EA of a new sewage works should cover associated infrastructure development such as roads, sewers, discharge pipes, and power supplies.

EA is a process which enables the developer and planning authorities to assess the various effects leading to the selection of a preferred treatment works process and location. It is summarized in the Environmental Statement (ES) which accompanies the Planning Application. In the UK, the developer is responsible for both the scope and the quality of the Environmental Statement. It is good practice to scope the EA in consultation with the statutory bodies such as the NRA, English Nature, the Countryside Commission, English Heritage, the Planning authorities, and the local community.

In the EA of sewage and sludge treatment, consideration is given to both significant operational and construction effects. The ES is a public document and offers an opportunity to describe how any potential negative effects will be resolved. Indirect effects and distant effects are also relevant. In sewage treatment, energy use may give rise to greater greenhouse gas production from the sludge.

Transport of solid waste will given rise to direct traffic impacts, both local and possibly distant, and indirect effects through fuel use (NO_x, CO_2, and particulate emissions). Effluent discharges also tend to have distant effects in terms of downstream water quality, which will be considered in the consenting process, but indirect water quality effects such as the risk of overflow from sewers due to stormflows, blockages, or pump failure also arise.

Site Selection

The primary aims of site selection for a waste management site are to ensure that the site conforms with the project specification and that the project is acceptable to the local community. This must be balanced with minimizing environmental impact and development costs.

For sewage treatment works, the discharge is usually the key environmental effect and therefore the location of the receiving water is a prime consideration. The site position in relation to the sewers is similarly critical. The process choice, discharge quality, and location, are therefore interrelated. Best results are achieved when water quality monitoring and modelling required for the consenting purposes is undertaken within the EA.

Table 8 : Typical constraints in sewage treatment site selection.

	Engineering	Environmental
Topography	Altitude-dependent on catchment and discharge heights *Slope*-dependent on process choice	*Visual*-low quality landscape or secluded
Population	Not within 400 m of residential development. Close to 'A' road for transport of sludge. Short sewage transfer time	Not protected heritage site. Not protected conservation site. Not area of outstanding natural beauty.
Receiving Waters	Close to receiving water. Not within flood plain. overflow route.	Avoid : bathing waters potable abstraction fisheries (salmonid) shellfishery sensitive areas (E)
SSSIs		
Socio-economic	Low land value (purchase price)	Low recreational value

In order to identify possible sites, an overlay technique can be used. The engineering and environmental constraints, which may be exclusionary or preferred, are specified (see Table 9).

Future land use and planning designations can be taken from local and county structure plans. The areas delimited by the constraints are drawn onto maps and overlaid. The resulting map will show areas most suitable for the sewage works construction. In practice, the site selection process may well rely on some compromise. For coastal towns the topography and earlier marine treatment concepts mean that the sewage usually drains through congested urban areas to the coast. The construction of the required sewage treatment facilities in urban locations poses significant environmental and engineering problems.

However, plants can be designed and constructed so that they blend into the local urban environment. Perhaps the biggest problem is to get the public to accept that a traditional wastewater treatment plant can become a 'purification centre' without the negative image that existing plants sometimes generate, Irrespective of whether plants are built above or below ground their outward appearance can completely disguise what is happening inside and, by careful design, no external smells or noise should occur. Careful choice of use of space above the plant can help offset high urban land costs. Building enclosed works is very expensive but so too is the construction of sewers and associated works needed to transfer sewage from the coast to a suitable inland site for treatment and return back to the coast.

Process Selection

The process selection cannot be entirely separated from the site selection. Once the potential sites are identified, and the likely discharge quality known, then the optimum process can be established for each site. Where landscape or odour and noise nuisance or land availability are significant considerations, compact treatment methods including biological aerated filtration (BAF), enhanced settlement, or high rate activated sludge offer benefits for covering and underground installation. However, for rural sites, conventional treatment offers energy efficiency and both cost and risk benefits. Reedbeds and lagoons, which are used more extensively in the USA and Europe, have low visual impact and are suitable in rural landscapes.

Industrial Effluents

Discharges to Sewer

Industrial effluent control is a highly regulated field. Regulation has driven considerable developments to minimize the impact of industrial waste on the environment. The principal legislation controlling industrial discharges to sewer is the Water Industry Act 1991. Trade effluent discharged to the public foul sewer must have the prior authorization of the Sewerage Undertaker in the form of a consent. This will include quantity and quality conditions and provides the framework of control.

Industrial effluents are inherently complex and variable. Effluents discharging into the sewerage system, alone or in combination, can result in gaseous emissions or blockages. Where the waste contains sulfur-containing organic compounds or sulfates, the activity of bacteria can form sulfides and other malodorous compounds. This occurs in anaerobic conditions such as rising mains or large low-turbulent gravity sewers and may lead to the formation and release of methane gas. The disposal of solvents is strictly controlled at source to prevent the formation of vapours, and the discharge of petrol to sewers is specifically prohibited. Accidental discharge of either could cause explosive atmospheres.

Industrial effluents need to be strictly controlled to ensure the build up of solids or deposition of grease does not cause a blockage which would result in the premature operation of stormwater overflows. The discharge of organic matter can significantly effect the ecological balance of the watercourse. The possible presence of toxic substances can also exacerbate the problem, although this should have been taken into consideration when setting the conditions on the sewer discharge consents.

Heavy metals such as zinc, copper, and lead are normally found in domestic sewage. Significantly higher concentrations of these and other metals are often found in industrial wastes. The majority of these metals end up in the resulting sewage sludge leading to problems if it is to be recycled to land. In all cases the industrial content of the sewage can substantially effect the final concentrations of toxic substances in the sludge for disposal. Control of metal bearing effluents at source is the prime method of reduction or elimination. This may require the industrialist to install an extensive pretreatment plant, such as metal precipitation, or the total containment of certain noxious wastes for disposal by a specialized contractor.

Certain industrial effluents can contain particularly persistent or toxic compounds such as cadmium, mercury, and persistent pesticides such as pentachlorophenol or γ-hexachlorocyclohexane. Some of these organic chemicals such as pentachlorophenol may undergo some biological degradation through the treatment process. A portion, however, will remain unaffected and retain its pesticidal properties. Its discharge into the receiving environment must be minimized at source. Release of pollutants must be prevented where possible or minimized, and all emissions must be rendered harmless to the environment. The controlling authority in England and Wales for these 'special category' substances is HMIP which applies the concept of Integrated Pollution Control (IPC) by the means of Best Practicable Environmental Option (BPEO). Many industrial processes result in the release of pollutants to a variety of receiving media and, under IPC, detailed limits and operating standards are imposed to control the range of emissions.

Some of the constituents of industrial effluents will remain in the liquid phase and may be discharged as part of the sewage works effluent. The NRA will set appropriate limits in the consent to protect the quality of the watercourse. Again in order to minimize the impact on the environment, certain industries may need targeting to reduce their inputs.

Discharges to Watercourses

Some industries opt to discharge direct to watercourses rather than to the public sewer system. Such discharges are controlled by the NRA who grant a consent to discharge similar to that which a Sewerage Undertaker grants to a sewer discharge. Conditions will be imposed which relate to the receiving watercourse using the same criteria used to determine the consent levels on a sewage works final effluent. Compared to a public foul sewer discharge, the conditions on the consent will be much tighter in view of the potentially immediate affect on the receiving water. The risk of failure is ever present and most industrialists prefer the option of a discharge to the sewage works. The sewage works, however, can be severely affected by a sudden strong industrial discharge. The treatment process may be inhibited with a deterioration in the final effluent quality and harm to the environment.

Direct treatment by the industrialist may prove difficult if there is an imbalance of available nutrients for conventional biological breakdown. The waste streams may lack nutrients such as nitrogen and phosphorus which will restrict biological activity. Mixing with domestic sewage at a sewage works will restore nutrient balance and allow the waste to be adequately treated. In these cases sewer disposal may be the best option. However, for some food manufacturing sites it may be preferable to discharge to sewer at the earliest opportunity and retain the site solely for food production to minimize any possible contamination of their product.

Other Waste Disposal

Whatever treatment process is chosen, the plant will inevitably produce other wastes for disposal to the environment in addition to the final liquid effluent. This can be solid waste which may be sent to landfill, such as pressed sludge, or a liquid sludge for off site disposal. Both are likely to involve haulage or tankering, with similar restrictions on disposal to those applied to sewage sludge. Metallic or complex organic compounds can be an environmental issue and must be considered when deciding on the optimum disposal route for the sludge. As before, BPEO will be applicable in such circumstances.

In some cases a solid or liquid waste produced as a consequence of an industrial process or effluent treatment will be designated a controlled waste. This type of waste will be subject to the duty of care laid down by the Environmental Protection Act 1990, the so-called 'cradle to the grave principle'. This ensures the security of the life history of a waste from the point of origin to the point of final disposal. This duty applies to the producer of the waste and requires that reasonable measures are taken to prevent the unauthorized or harmful deposit, treatment, or disposal of the waste and to prevent the escape of waste from his or any other person's control. In addition, the producer must ensure that the transfer of waste is made only by an authorized person and with appropriate documentation.

Effect of UWWTD

The UWWTD will have a significant impact on industrial effluent disposal. The introduction of a biological secondary treatment stage in addition to the solids separation primary stage at some esturial and coastal sewage works will significantly increase industrial sewage disposal costs. The Directive also requires that certain industrial effluents which discharge directly to the aquatic environment will have to conform to particular standards which will be determined on an individual case basis by the appropriate regulatory authority. This applies to biologically degradable wastes with a population equivalent of > 4000. These industries, chiefly the food processing sector, will need to extensively review their current effluent production and disposal facilities. In future, the implementation of waste reduction and effluent management programmes, both for direct river discharges and sewer disposal, are inevitable to further reduce the impact on the environment.

Future Developments

Factors Influencing Direction

The improvements and new technologies that will emerge for sewage and industrial effluent treatment into the next century will largely be determined by the driving forces influencing the industries over the next few years.

In the European Union and much of the developed world the main factors are legislation, costs, and public image. In addition, land area requirements or water re-use are strong influences in some countries where space is at a premium (such as Japan) or water supply is limited (as in Australia and parts of the USA). In developing countries the main driving forces influencing development of sewage treatment are usually costs, the need for basic health care, and lack of water.

The implementation of legislation in some third world countries is often haphazard, with relatively little regard for public concerns. There is, however, a desperate need for appropriate basic low-cost sewage treatment, much of which is already available if it can be afforded and has the commitment of governments.

In the industrialized nations public image and perception often leads to new legislation. This, in turn, leads to costs in implementation which then create a new public image. This is clearly the case where concern over environmental factors has been partly replaced by concern over the cost to the public in implementation. However, the trend towards higher standards of treatment whilst making the most efficient usage of existing assets will inevitably continue.

Any new technologies and process improvements will be influenced by a variety of factors including climate, geography, and–more importantly–the investment in existing assets. Improvements to, or optimization of, existing processes will be dominant as there are relatively few instances where complete new treatment works will be needed.

When limited areas of land are available in urban locations there will have to be a choice between pumping the wastewater inland to a conventional treatment works or constructing a covered urban installation. The implementation of either type of solution will be a major influence on the locality and it will be essential to initiate early full environmental assessments together with full public consultation.

In both Europe and Japan there are examples where local circumstances have led to the provision of office buildings, car parks, gardens, or sporting facilities as part of urban sewage treatment works. In some Japanese cities new buildings over a certain size must now incorporate sewage treatment/greywater recycling systems.

Although there is sufficient knov'ledge to construct and operate covered or underground installations safely, and to prevent noise and odour nuisance, it is unlikely that sewage treatment will readily lose the negative image which originates with the existing traditional plants.

The possibility of constructing properly designed treatment plants in the midst of residential areas will require a comprehensive change in public opinion over may years. In larger communities it may be that a better alternative to a town centre development would be to pump the sewage to a suburban industrial area or manufacturing zone where the provision of a treatment works as a single-storey 'purification-factory' would be more acceptable.

An additional advantage of this alternative is that the few town-centre sites which are available remain for use as less controversial underground storm water settlement and attenuation tanks which will also be needed in the future. Fortunately, during the last decade new treatment methods have been developed which will enable secondary treatment to be provided on very much smaller sites than traditional processes. Table 9 compares relative land area requirement for different treatment processes. Extra treatment capacity and land is required for works which nitrify the ammonia content than for those performing only carbonaceous oxidation.

Table 9 : Relative land areas required for sewage treatment.

Process	Nitrifying/%	Non-nitrifying/%
Biological filters	100	50
Activated Sludge	12	7
Biological Aerated Filter	2	1
Membrane Technology	6	5
Chemical Dosing, etc.	80	40

Process Optimization

As has been discussed earlier, there is a wide range of different components to be found at treatment works. With this comes an even greater number of ways of operating them. In order to optimize the operation of treatment processes and select the optimum operating conditions for individual process components, increasing use of computer modelling and process simulation will be made.

The use of computer software in process simulation of works performance allows the optimization of the operation of the whole works to give higher standards of treatment and more efficient usage of existing assets. Such use of software makes it relatively easy to evaluate performance under different conditions without the expense or risk associated with running real trials on a functioning works. Additionally, computer modelling of specific components such as settlement tanks and aeration lanes will lead to modifications in design and improvements in treatment unit performance.

In conjunction with modelling and practical trials, a number of broad strategies will be examined to improve commercial performance in the water industry and these are given in Table 10.

Process Integration

Where feasible, the integration of sewage treatment processes with other industries offers significant advantages. The siting of a power station near industrial processing units and a sewage works allows good quality effluent to be provided for power station cooling water which, in turn, supplies excess heat to industrial and food processing units; these, in turn, produce effluent for treatment. When energy from biogas production and soil conditioners/fertilizers from sludge drying processes are added, not only are treatment costs subsidized, but the integrated 'environmentally sound' approach will improve public image.

Table 10 : Process optimization.

Flow/load balancing	Work in the USA indicates that up to 15% additional treatment capacity could be achieved.
On-line monitoring	Diversion of incoming feed to balancing tanks when monitors indicate to control shock and peak loads. Diversion of effluent for 'polishing' when instrumentation indicates it to be required.
Provide load management within works	Avoid operationally generated shock loads, provide small separate treatment devices within the works to contain shock loads and protect the 'core' process, *e.g.* mini-filter on supernatant return.
Pre-treatment in sewers/rising mains	Dependent on appropriate locations. Modern instrumentation offers scope for additional control both within the 'treatment pipe' and within the catchment.
Chemical dosing	Greater use of chemical treatment to up-rate primary treatment and thus decrease loadings on secondary

	treatment. Economic analysis will be used to determine if higher running costs are offset by capital savings. Use of dosing to control peak loadings to works.
Review loading rates	Analysis indicates standard design parameters are frequently conservative compared to actual results. New loading rates should reflect the required method of monitoring. Settlement rates can be enhanced by lamella type devices or by chemical dosing, but this must be accompanied by a review of desludging and sludge thickening practice. Biochemical loading rates could be enhanced by better distribution, insulation and weather protection of units, recirculation of effluent, and flow balancing.
Review traditional	The enhancement of existing assets will include
design and operational practice	improved media in existing biological filters, improved larger aerators in existing tanks, revised operating sequences, and alternative desludging regimes.
Improve 'critical' facets	Examine the benefits of covering biological filters,
of treatment processes	pumped recirculation, regular and controlled filter 'flushing', and settlement enhancement devices
Tighter trade control	Minimize input of difficult wastes to the works. Improve control of peak loads to works. Greater use of on-site treatment to deal with low volume high strength trade waste at source as an alternative to extension of works.

New Technologies

To date, the introduction of new technologies into sewage treatment has been slow. In part this is due to the conservative nature of an industry where processes are expected to last in excess of 20 years. This, coupled to the high cost of introducing new technology to the treatment of a low value product has lead to few truly new systems being introduced. However, as costs for new technology reduce and the need to improve treatment and process economics increases, this situation will change. Existing processes have been highlighted earlier, but a few of the new techniques that will influence future developments are mentioned here.

On-line Monitoring

The introduction of new and more reliable sensors and on-line monitors for ammonia, COD, respiration rate, turbidity, and specific pollutants, allows treatment processes to be controlled in real time. This allows a greater control of effluent quality and improves process economics, by allowing the treatment plant to be operated closer to the margins and reduces the need to design in some overcapacity.

Radiation Pulse Treatment

Use of high voltage pulses have been reported to give disinfection, destruction of soluble organics, and rapid dewatering of sludges. The combination of shock wave and soft X-rays could offer significant advantages over conventional treatment, although public acceptance of the technique and its costs have yet to be determined.

Membrane Separation Systems

Recently a number of sewage treatment systems have been developed utilizing membrane filtration. Originally developed for food and pharmaceutical industries, membrane systems have now seen widespread large-scale application in desalination and water treatment plants. This technology relies on a microporous barrier (normally polymeric) in order to filter effluent on the basis of particle size (usually to < 1 μm). To reduce fouling, membranes are generally used in a cross-flow arrangement, with the bulk of the effluent flow across the membrane surface through which treated permeate is removed. The process has the advantage of being able to deliver disinfected effluent of uniformly high quality, but can have a significant cost disadvantage compared with conventional treatment. Nonetheless, as perhaps the only truly new sewage treatment process to have been developed in the last ten years, membrane systems have attracted considerable interest from water companies.

Two specific developments are worthy of mention:

(i) use of membrane separation as an alternative to biological secondary treatment, followed by disinfection for coastal sites, and

(ii) membrane bioreactor systems.

Membrane Secondary Treatment. Screening and primary treatment is followed by membrane separation to remove remaining suspended solids and achieve disinfection by physical removal of pathogenic organisms, including bacteria and viruses. Since discharge of soluble BOD is rarely an issue at coastal sites, membrane treatment can significantly reduce capital expenditure as biological treatment, settlement, disinfection, and outfall costs are decreased. However, operating costs are higher and overall cost effectiveness depends on site-specific factors. Of the systems available, those from Memcor and Renovexx are now seeing full scale application.

Membrane Bioreactor Systems. Membrane bioreactors combine biological treatment and barrier separation stages in one treatment system. Use of membranes to replace settlement tanks allows the retention of high levels of biomass in the biological treatment stage (typically 15 to 20 g l^{-1} mixed liquor suspended solids). This provides a compact system, giving a very high quality disinfected effluent and lower waste sludge volumes due to thicker sludges than conventional processes. However, the current cost of membranes limits cost competitiveness to smaller scale systems where high effluent quality or restricted land area is a concern. The fully automated system developed by Kubota, in Japan, probably represents the forefront in advanced sewage treatment systems.

Secondary (Biological) Treatment Systems in more Detail

Biological treatment is the removal of biodegradable organic materials (expressed as BOD) from water (sewage) and often includes biological oxidation of ammonia to nitrate, sometimes biological reduction of nitrate to nitrogen (a gas which escapes to atmosphere during aeration) and occasionally biological removal of phosphates (arising from detergents) when required. The oxidation of ammonia to nitrate is called nitrification and the reduction of nitrate to nitrogen is called denitrification.

Organic materials are compounds of carbon and the microorganisms that remove organic materials are called carbonaceous microorganisms. These occur naturally in the environment. Ammonia (NH_3) and nitrate (NO_3) are compounds of nitrogen (N) and the microorganisms that oxidize ammonia are called nitrifying bacteria. Nitrifying bacteria do not have a significant function in the removal of organics.

If part of the plant is modified to suit, some of the carbonaceous microorganisms can remove nitrate and phosphate. To effect purification as quickly as possible the concentration of microorganisms in the sewage must be increased substantially. For aerobic microorganisms to multiply quickly they need plentiful supplies of food (biodegradable organics) and dissolved oxygen. The organics are already in the sewage and therefore their only requirement is oxygen. Sewage-treatment plant has an aeration facility for this purpose.

Activated-sludge Systems

These are a range of biological treatment processes in which the microorganisms are grown as a free suspension in water which appears in the aeration tank as a rich brown slurry called activated sludge. Microorganisms present in the sewage multiply rapidly in the aeration tank due to the mixing and oxygenation (aeration). The mixture of treated sewage plus microorganisms (mixed liquor) flows from the aeration tank to the settlement tank where the microorganisms (activated sludge) settle to the bottom and treated sewage (secondary effluent) overflows a weir. Most of the settled activated sludge is pumped back to the aeration tank to treat more sewage and the extra growth of sludge, which is not required, is removed from the plant (excess sludge).

The conventional activated-sludge process requires primary settlement (before biological treatment) which, in a two-hour retention period, can achieve 30 per cent BOD removal and 60 per cent suspended-solids removal. The remaining dissolved, colloidal and fine suspended organics can be removed in a short aeration period. However the extended aeration modification to the process does not use primary treatment and the period of aeration needs to be considerably extended to be able to accept the whole organic load in the screened sewage.

It would be normal to have average retention periods in aeration for conventional activated sludge and extended aeration of about 6 hours and 24 hours respectively, although these values are not rigid or accurate but depend on the strength of the sewage, the effluent quality required and any consideration of quality and stability of excess sludge.

Contact stabilization is a form of activated-sludge process that attempts to achieve a given effluent quality with smaller aeration tanks. It uses the ability of activated sludge to rapidly absorb organics (BOD) after the sludge has been aerated without food for several hours. Contact stabilization is considered less reliable than conventional activated sludge or extended aeration in producing a good effluent quality. This is due to factors such as variations in sewage flow creating variations in retention time in the contact tank. Unlike other processes if the retention time is either too long or too short in the contact tank, the effluent quality will deteriorate.

Biological filters, sometimes called trickling or percolating filters or filter beds, are not filters in the normal sense, that is they do not remove suspended solids by passing the flow through a porous medium. This is a biological contact process of the fixed-film or attached-growth type and, like the activated-sludge process, uses natural aerobic microorganisms to remove biodegradable organic material and, when required, ammonia from sewage. The major differences from the activated-sludge process are:

- the microorganisms grow as a film or slime on the surface of coarse media (traditionally 50mm stones or, more recently, special plastic media);
- the dissolved oxygen is obtained by natural convection of air through the filter; and
- because the microorganisms are fixed, no sludge return is necessary although secondary sedimentation is required to remove the excess microorganisms (humus sludge) that are discharged from the filter.

The depth of the medium is traditionally about 2m although newer designs using effluent recycling can have much greater depth.

Circular filters have rotating sewage distributors driven by the flow of sewage but the distributor can be electrically driven, as is usual for rectangular filters. Coarse suspended solids can choke the filter and therefore primary treatment always precedes the filter.

Rotating Biological Contactors (RBC)

This process is based on the principle that when an object is rotated about a horizontal axis, partly immersed in sewage, it first becomes wetted and later begins to grow a layer of microorganisms which feed on the organics in the sewage and obtain dissolved oxygen that transfers from the air to the wet film.

The RBC can be a series of discs, a continuous spiral or open plastic matrix. The biological film grows to a thickness dependent on organic load and the speed of rotation. Further growth of film is flushed from the biological contactor into the effluent flow and can be removed by settlement or fine screens.

RBCs require primary treatment (before biological treatment) because they cannot remove suspended solids and the mixing in the RBC chamber is not sufficient to prevent the deposition of primary sludge. The finer suspended solids leaving primary treatment flocculate into larger particles in the RBC chamber and settle with the excess biological solids to form a secondary sludge in the secondary settlement tanks.

The free spaces within the RBC must be large enough to prevent adjacent biological films from meeting and choking the contactor. This is also dependent on the speed of rotation which is usually between *1 and 3 rpm.*

It is advisable for the RBC tank to be divided into at least 3 stages, sometimes 5 stages, to achieve high-quality effluents. The total contact area required is dependent on the amount and proportion of BOD and ammonia to be removed. The RBC is usually covered to prevent the biological film being stripped off by wind or rain, and to prevent deposition of wind-blown sand.

Large plants have primary and secondary settlement tanks whereas small-package plants usually have pretreatment by septic tank, often including flow equalization, plus a separate secondary settlement chamber or drum filter (microscreen) with return of screened sludge to septic tank.

Small Treatment Works

For low population application, of the order of one thousand down to individual households, treatment economics dictate a more simplified approach.

The normal functions of an inlet works are flow measurement, screening and grit removal; the latter two processes are principally to protect downstream pipe-lines and units from blockage.

Generally, on very small plants flow measurement can be considered as being an unnecessary complication, whilst grit removal should only be contemplated if there is particular reason to suspect high grit concentrations from the sewers.

The need for screening is usually more obvious, although the practice should be avoided as far as possible simply because if screens are installed they have to be cleaned. In-line or pump-suction maceration can be a cost-effective alternative.

Septic tanks provide a useful facility for the treatment of sewage on small plants. It should be stressed that cesspools are not treatment units but only storage vessels and now too rarely justified to require consideration.

Multiple-compartment tanks are usually favoured for populations above 30, when about two-thirds of the capacity is provided in the first compartment. For installations serving over about 60 people, there are advantages in having two separate tanks operating in duplicate.

The arrangement for internal baffles at inlet and outlet are important, since obviously the introduction of crude sewage and the removal of clarified liquid should cause the least possible disturbance of the settled sludge and the surface scum. There are good designs for prefabricated units, either constructed virtually at site from concrete pipes or bought to site ready made in glass-reinforced plastic.

The case for providing septic tanks in two separate parts has been pressed and is perhaps strongest where there is any doubt regarding the regularity or reliability of the tank management and sludge removal.

The second tank has significant advantage as a 'catch fence' and should be adopted generally for individual tanks serving 12-100 persons.

The effluent from a small septic tank is most commonly dealt with by discharge to a soakaway system either in the form of a hardcore filled pit or as the size increases by percolation through a field-drain system. The permeability of the soil is most relevant to the system and should be assessed or tested before the system is sized.

Table 11 : Design of small treatment works: relative demands of power, land, capital and operator requirement.

Process	Land	Capital cost	Power consumption	Operator demands
Activated sludge	1.0	1.00	1.00	1.0
Extended aeration	1.2	1.16	2.20	0.5
Biological filter	4.5	1.83	0.40	0.7
Rotating contactor	1.6	2.22	0.50	0.4
Oxidation ponds	78.0	1.07	0.08	0.1

Due to their complexity, conventional activated-sludge systems are not normally applied to small treatment works. However, modifications of the system are most effective and can be applied in small plants to unsettled crude sewage. The common processes are usually contact stabilization or extended aeration systems, and these processes use a variety of aeration devices to mix and aerate the mixed liquor. Both systems have the ability to absorb shock pollutant loads but the simplicity of the extended aeration system has given it prominence.

Packaged rotating biological contactor units are becoming increasingly popular due to the low energy requirement, low profile, low noise level and ease of full enclosure.